THE LEGEND OF THE TITAN CORPORATION

JEFFREY L. RODENGEN
AND
RICHARD F. HUBBARD

Edited by Melody Maysonet
Design and layout by Sandy Cruz

Write Stuff Enterprises, Inc.
1001 South Andrews Avenue
Second Floor
Fort Lauderdale, FL 33316
1-800-900-Book (1-800-900-2665)
(954) 462-6657
www.writestuffbooks.com

Copyright © 2002 by Write Stuff Enterprises, Inc. All rights reserved. No part of this book may be reproduced or transmitted in any form by any means, electronic or mechanical, including photocopying and recording, or by any information storage or retrieval system, without permission in writing from the publisher.

Publisher's Cataloging in Publication

Rodengen, Jeffrey L.
 The legend of the Titan Corporation/ Jeffrey L. Rodengen. & Richard F. Hubbard; edited by Melody Maysonet; design and layout by Sandy Cruz — 1st ed.
 p. cm.
 Includes bibliographical references and index.
 LCCN 2001135148
 ISBN 0-945903-74-X

 1. Titan Corporation—History.
 2. Telecommunication equipment industry—United States. 3. Defense industries—United States. I. Hubbard, Richard F. II. Title.

 HD9696.T444T58 2002   384'.06573
                          QBI02-200674

Completely produced in the
United States of America
10 9 8 7 6 5 4 3 2 1

Also by Jeffrey L. Rodengen

The Legend of Chris-Craft

IRON FIST: The Lives of Carl Kiekhaefer

Evinrude-Johnson and The Legend of OMC

Serving the Silent Service: The Legend of Electric Boat

The Legend of Dr Pepper/Seven-Up

The Legend of Honeywell

The Legend of Briggs & Stratton

The Legend of Ingersoll-Rand

The Legend of Stanley: 150 Years of The Stanley Works

The MicroAge Way

The Legend of Halliburton

The Legend of York International

The Legend of Nucor Corporation

The Legend of Goodyear: The First 100 Years

The Legend of AMP

The Legend of Cessna

The Legend of VF Corporation

The Spirit of AMD

The Legend of Rowan

New Horizons: The Story of Ashland Inc.

The History of American Standard

The Legend of Mercury Marine

The Legend of Federal-Mogul

Against the Odds: Inter-Tel—The First 30 Years

The Legend of Pfizer

State of the Heart: The Practical Guide to Your Heart and Heart Surgery
with Larry W. Stephenson, M.D.

The Legend of Worthington Industries

The Legend of Trinity Industries, Inc.

The Legend of IBP, Inc.

The Legend of Cornelius Vanderbilt Whitney

The Legend of Amdahl

The Legend of Litton Industries

The Legend of Gulfstream

The Legend of Bertram
with David A. Patten

The Legend of Ritchie Bros. Auctioneers

The Legend of ALLTEL
with David A. Patten

The Yes, you can of Invacare Corporation
with Anthony L. Wall

The Ship in the Balloon: The Story of Boston Scientific and the Development of Less-Invasive Medicine

The Legend of Day & Zimmermann

The Legend of Noble Drilling

Fifty Years of Innovation: Kulicke & Soffa

Biomet—From Warsaw to the World
with Richard F. Hubbard

NRA: An American Legend

The Heritage and Values of RPM, Inc.

The Marmon Group: The First Fifty Years

The Legend of Grainger

TABLE OF CONTENTS

Introduction vi

Acknowledgments viii

Chapter I   Starting Small 10

Chapter II   The Early Years 22

Chapter III Merging into the Fast Lane 30

Chapter IV   A Shifting World 44

Chapter V   Meeting the Challenge 54

Chapter VI   Innovation and Entrepreneurship 64

Chapter VII Creating Value 74

Chapter VIII Maximizing Value 86

Chapter IX   Titan to the Rescue 100

Notes to Sources 132

Index 138

Introduction

THE LEGEND OF THE TITAN Corporation is a classic American success story—a story of how hard work, skillful maneuvering, and a willingness to try new ideas turned a dream into reality. Founded in 1981 by Gene Ray, Ed Knauf, and Jack McDougall, Titan started as a defense contractor during the Cold War, when military spending was on the rise. The company started small but soon turned into a powerhouse that lived up to its ambitious name.

From the beginning, the founders shared ownership of Titan with employees and fostered a cooperative rather than a competitive environment. As a result, they were able to recruit a top-notch staff early on and quickly build an attractive reputation in the defense technology industry.

Owing to President Reagan's Strategic Defense Initiative (SDI), or "Star Wars" missile defense program, Titan could barely keep up with opportunities for government defense-related contracts. As early as 1983, Titan began acquiring companies that enhanced its expertise in advanced technology, system development and integration, system analysis, and specialized products. Then in 1985 it became a publicly traded corporation when it merged with Electronic Memories and Magnetics (EMM), a manufacturer of computer and memory products more than three times the size of Titan. Sidney Webb, EMM's chairman, brought his sharp business sense to Titan as its new chairman, and Gene Ray became president and CEO. Though Titan divested many of EMM's businesses, the merger opened up commercial markets, a windfall that would prove extremely important.

Ray and Webb began adding a cadre of new talent to the already gifted pool of employees and directors, men and women who were more than happy to join the Titan culture of innovation, flexibility, and fun. More acquisitions followed, each one carefully evaluated to ensure a strategic and cultural fit, and the company streamlined to better serve its growing list of customers.

By 1987, the Cold War was drawing to an end, but the longed-for peace had a serious impact on the defense industry. The stock price of defense companies began a downward slide, and though much of Titan's business was involved with C^3I (command, control, communications, and intelligence) contracts, it couldn't escape the "taint of the Star Wars brush." As the government pulled back on defense spending, Ray and Webb shrewdly moved Titan's technologies into commercial markets.

Though a backlog of contracts carried Titan through the worst of the slump, the company began some serious restructuring and cost-cutting. At the same time, it made strategic acquisitions to round out its project capabilities and put a sharper focus on C³I technology—one of the few defense segments that escaped the Pentagon's budget cuts. Titan also began to diversify by exploring new technologies with potential commercial application. By the time the Cold War ended in 1989, Titan's precarious position had stabilized somewhat.

Ray, Webb, and Titan's other leaders still faced the greatest test in the company's history as the defense budget continued to shrink and many contracts were canceled. Then in 1990 Titan's commercial segment got a big boost when it acquired Linkabit Corporation, a manufacturer of earth-based satellite communications equipment for the U.S. government and the maker of Mini-DAMA, a sophisticated satellite communications terminal. Other strategic acquisitions and partnerships followed, as did new nonmilitary and military contracts.

In 1992 Titan made a pioneering move by creating a new commercial business from an existing technology. Titan's patented electron beam sterilization system—later called SureBeam®—killed bacteria on medical devices and provided an alternative to existing systems that used radioactive materials or hazardous chemicals. By the end of the year, Titan's willingness to try new ideas had paid off: It was one of the few technology companies that had found success in diversification. Meanwhile, Titan continued to focus on its core businesses of communications and information systems and services and made strategic acquisitions in defense technology companies.

In 1997, Ray hired Eric DeMarco as CFO and senior vice president. Ray and DeMarco realigned Titan into focused companies and planned to grow each business profitably and efficiently while maximizing shareholder value through spin-outs and spin-offs.

The new strategies paid off rapidly and well, and Titan went on an acquisition spree. Its stock skyrocketed in 1999, and in 2000 Titan became a billion-dollar corporation.

Today, Titan is a diversified corporation with a well-rounded balance of military, government, and civil contracts. Titan Systems supplies most of Titan's new technology and is a leader in providing information systems solutions and services to the U.S. government. Even before the terrorist attacks of September 11, 2001, world events had created a need for more-sophisticated technology to protect the country, and Titan Systems met the challenge. SureBeam, now an independent, publicly traded company, has also performed exceptionally well, especially since the FDA and USDA approved the use of electron beam irradiation to kill harmful pathogens in meat and other foods. Titan has found many other uses for SureBeam, including irradiation of the U.S. mail to destroy anthrax. Cayenta, the company's e-business solutions subsidiary, grew substantially in only a few years, but in 2002 Titan decided to sell certain of its operations to focus on the utility and municipality market. Titan Wireless performed exceedingly well until 2002, when it faced the deteriorating telecommunications market, and Titan made the difficult decision to sell the business. As part of its continuing strategy to create new commercial businesses from existing technologies, Titan has also developed E-tenna, LinCom Wireless, and Titan Scan, each with its own leadership and employee stock ownership plan.

The quality of Titan's leadership, headed by Gene Ray, chairman and CEO, and Eric DeMarco, who became president and COO in 2002, has set it apart, and the company's ongoing innovative technological advances in communications and information systems have exposed the world to future benefits and positioned the company to take advantage of future opportunities. Titan's success at creating new technologies and finding new markets has enabled it to survive and thrive in an ever changing marketplace and an ever changing world.

Acknowledgments

A GREAT NUMBER OF people assisted in the research, preparation, and publication of *The Legend of The Titan Corporation*.

The principal research and narrative time line were the work of our research assistant Barbara Fitzsimmons, whose efforts went a long way toward making this book a success.

This book would not have been possible without the vivid and forthright recollections of Gene Ray, chairman and CEO. Ralph "Wil" Williams, vice president of corporate communications, was also instrumental in the project's development. Both men lent their tireless support to see the book's completion.

Many other Titan executives, employees, retirees, directors, and associates greatly enriched the book by discussing their experiences at or with the company. The authors extend particular gratitude to these men and women for their candid recollections and anecdotes: Thomas Allen, vice president of systems engineering integration; John Arme, SureBeam board member; the late M. C. "Bud" Baird, president and CEO of Titan Systems; Edward Bersoff, board member; Rochelle Bold, vice president of investor relations; Ray Calhoun, former director of sales for SureBeam; Joseph Caligiuri, board member; Mickey Copeland, retired executive assistant to Gene Ray; Congressman Randy Cunningham; Eric DeMarco, president and chief operating officer; Rolf Erikson, retired vice president of corporate development; Daniel Fink, board member; Tony Frederickson, president of Titan Systems' Applied Technology Group; Andrew Gaspar, financial analyst; Victor Gogolak, retired vice president; Ron Gorda, president of Titan Systems' Information Products Group; Karl Gould, retired vice president of Titan Systems; Congressman Duncan Hunter; Ron Jacobson, senior vice president of advanced technologies for Linkabit; Nancy Jenkins, former assistant to Ed Knauf; Ted Kavanaugh, retired CFO and senior vice president; Linda Frady Keenan, director of contracts for Titan Systems; Albert "Ed" Knauf, cofounder of Titan and retired president of Titan Systems; Ken Kreyenhagen, retired vice president of research and technology; Michael Kulinski, former COO and chief technology officer of Titan Wireless; Richard Llewellyn, retired vice president; Gary Loda, retired president of the Beta Development division; John "Jack" McDougall, Titan cofounder and retired executive vice president; Marshall Nelson, retired general counsel, secretary, and senior vice president; Edgar Northrup, retired senior vice president; Larry Oberkfell, president and CEO of SureBeam; Virginia Oliver, assistant corporate controller; Denny Olson, vice president of food technology for SureBeam; Cathy Pergam-Ball, retired vice president; Earl Pontius, president of the technical resources sector of Titan Systems; David Porreca, retired president and CEO of Cayenta; Charles "Chuck" Saffell, president of the maritime sector of Titan Systems; Joseph Saponaro, president of the civil sector of Titan Systems; Diane Scott, vice president of human resources; Philip Spence, operations manager of the Pulse Sciences division of Titan Systems; Mary Squazzo, executive assistant to Gene Ray; Thomas Trimble, senior vice president of Linkabit; Brian Williams, principal technical advisor to SureBeam; and Bill Zettinger, vice president of programs for Linkabit.

As always, special thanks are extended to the dedicated staff at Write Stuff Enterprises, Inc.: Jon VanZile, executive editor; Melody Maysonet, senior editor; Heather Deeley, associate editor; Bonnie Freeman, copyeditor; Mary Aaron, transcriptionist; Barbara Koch, indexer; Sandy Cruz, senior art director; Rachelle Donley, Wendy Iverson, and Dennis Shockley, art directors; Bruce Borich, production manager; Marianne Roberts, vice president of administration; Sherry Hasso, bookkeeper; Linda Edell, executive assistant to the author; Lars Jessen, director of worldwide marketing; Joel Colby, sales and promotions manager; Rory Schmer, distribution supervisor; and Jennifer Walter, administrative assistant.

During the Vietnam War, Gene Ray, cofounder of Titan, did high-level work for the Department of Defense. In one of his studies, Dr. Ray used the best simulation technology available to determine that the U.S. Air Force should fly its B-52s at a high altitude for certain targets rather than the low altitude that had previously been determined. The results of Ray's analysis caused quite a stir, but in the end the exactitude of his work spoke for itself. The Air Force followed his recommendation, which helped protect the B-52s from being shot down.

CHAPTER ONE
STARTING SMALL
1981

My feeling was that if the three of us couldn't make the company go, then nobody could.

—Gene Ray, 1983

TO THE UNSUSPECTING eye, The Titan Corporation's beginnings in 1981 might have appeared unpromising. The initial business plan was scribbled on cocktail napkins; its initial capitalization came from personal savings and a small bank loan. The first headquarters were the three cofounders' homes. But when Gene W. Ray, Albert E. Knauf, and John R. McDougall chose the name "TITAN Systems" for their business (at first using all capital letters for Titan), they were planning nothing less than a powerhouse of industry. They calculated they could apply their scientific knowledge and first-rate military contacts to build a business that would earn millions, and eventually billions, of dollars.

Today, their dreams have become reality. Yet the success was not achieved without years of hard work, skillful maneuvering around fluctuating national defense budgets, and an extraordinary willingness to try new ideas.

Titan's home city, San Diego, has been a stronghold of military-related industries. But over time, many of San Diego's defense businesses have shrunk or disappeared. As the nation transitioned out of the Cold War and out of heavy defense spending during the 1980s and 1990s, many defense-related businesses were unable to adapt.

Titan, on the other hand, grew larger by finding niches in the defense industry, performing superbly, and venturing into new business arenas—some defense related, some not. The business that was started on napkins overcame formidable economic odds. Today it is stepping into a bold new future on several exciting frontiers.

A Powerhouse Is Born

The Titan story began in April 1981 at a restaurant in Alexandria, Virginia. Gathered around the table were three people: Gene Ray, a brilliant former Pentagon physicist who had also proven himself a savvy businessman through his work at San Diego–based Science Applications Inc. (later called Science Applications International Corporation, or SAIC); Albert "Ed" Knauf Jr., a former SAIC employee of Ray's who also had high-level military experience and had worked with Ray at the Pentagon; and Sue Knauf, wife of Ed.

Ray and Ed Knauf had already agreed to become equal partners in a start-up company, but it was during the April meeting at the restaurant that they actually put their business dreams on paper—on

---

Gene Ray spent his early years on a Kentucky farm that lacked electricity and running water. Finances were tight for the Rays, but there was never any question that Gene would attend college.

napkins, more specifically. The two men excitedly scribbled down their ideas while Sue looked on with some anxiety. "They were like kids in a candy store," she remembered. "I was absolutely sure we were headed for bankruptcy."[1]

That day, Ray and Knauf's projections spilled from napkins onto the tablecloth. Shortly thereafter, the entrepreneurs moved their planning to the Ray home, near San Diego, and the Knauf home, in Vienna, Virginia, just outside Washington, D.C., and set up shop with a typewriter, a telecopier, and a photocopier.

John "Jack" McDougall, former corporate vice president and manager of the Systems Group for SAIC, was Titan's third equal partner, "the partner to keep Gene and me in line," said Knauf, "because we were pretty strong-willed guys."[2]

McDougall had a master's degree in nuclear engineering from the University of Washington and had spent nine years in nuclear weapons effects at advanced-technology giant TRW before going to work for Gene Ray at SAIC in 1972. When McDougall told Ray in 1981 that he was going to resign from SAIC, Ray invited him to join in starting a new company.[3]

There was one question all three businessmen had to face early on: Did San Diego need another defense contractor? Certainly not to duplicate the achievements of giant General Dynamics or the work of upstart SAIC, where all three men had worked. In fact, Ray had signed a noncompete agreement with SAIC. The three partners determined that Titan would have to fill different defense needs.

The nation was in the midst of a recession, but defense spending was on the rise, and Ray, Knauf, and McDougall felt the business could succeed. Their strategy was to model the company after TRW, as "a systems engineering, systems integration, software and technology company." In an interview with the *Los Angeles Times*, Ray said, "That means we're not actually bending tin, but we're taking tin that somebody else bent and integrating it together and adding more value to it."[4]

Or, as Ed Knauf said 20 years after Titan's founding, "The basic, core strategy was to build a service company to generate the cash. Then the value of our stock would allow us to acquire technology companies that could get us into the product area."[5]

The business focus could be summed up in a word: communications.

Titan Systems' initial capitalization was about $200,000. The partners also quickly established a $500,000 line of credit with Bank of America, using their homes and additional stock holdings as collateral. There was a lot at stake, and Ray, Knauf, and McDougall were nervous. Soon there would be even more on the line.

At Horizons Technology, the company where Knauf had worked immediately before joining the new Titan venture, there was concern that some employees would follow Knauf out the door—and rightly so. Knauf had already signed up some of his Horizons people to start in a few weeks. Horizons' president called a meeting with those employees, who had independently and quietly planned to hand in their resignations, and asked them directly whether they planned to follow Knauf. When they said yes, they were told to leave.

"We were sitting on the curb with our boxes, looking at each other, and saying, 'What do we do now?'" recalled Richard Llewellyn, one of the victims of the ouster. Llewellyn had planned on at least two weeks of transition before actually joining the new company. "So I called Gene Ray up and told him what had happened. Gene said, 'Well, this is not quite what we had planned, but sure, come on board.'"[6] (As an interesting sidenote, Titan would one day acquire Horizons.)

Titan had no contracts and no business, but it had three new employees: Cathy Pergam, Llewellyn, and Victor Gogolak. Fortunately, those new employees were go-getters.

"Gene and I looked at each other and said, 'How the hell are we going to afford these guys?'" Knauf recalled. (At the time, Ray was writing payroll checks from his personal checkbook.) "We answered our own question: 'I guess we'll find out.'"[7]

While Ray and McDougall were working out of their homes in San Diego, the other employees were working out of Knauf's basement on the East Coast. "The space that Gene and Ed had leased wasn't ready yet," said Vic Gogolak. "So we wound up meeting in Ed Knauf's basement, which had one of the smallest dogs, who was one of the biggest challenges. We had to corral this dog to keep him from eating the proverbial homework. We were writing proposals, and all of a sudden this

One of Titan's first contracts involved an analysis of a new missile system called the MX.

little white thing would run through, putting paw prints on everything."[8]

"Vic and Richard and I would go to work in Ed's house every morning," Cathy Pergam (later Cathy Pergam-Ball) remembered. "One of us would bring the doughnuts. And after Ed ate about half the box of doughnuts, he'd go out and look for work for all of us. That would leave Vic and Richard and me to sit around to figure out what to do for the day."[9]

Pergam-Ball remembered with some humor how Titan Systems won its first competitive contract. "We were sitting around, and a friend of mine called me to ask if we were bidding on a contract to do MX studies." The friend had read about the study in *Commerce Business Daily*. "So I scurried around and got a copy of the *CBD*, and sure enough, there it was." Pergam-Ball asked the contractor if Titan could have until Monday to submit a proposal. The man she spoke to was doubtful that Titan could start and finish a proposal during one weekend but agreed to give the new company a chance.[10]

"So all that weekend we wrote a proposal for GTE Sylvania related to the MX missile system," Knauf recalled. "Rich Llewellyn and I flew to Boston on Monday and came back with a contract for $414,000.'"[11]

Titan was on its way. By its second month, the company had reached profitability, and new contracts rolled in at a steady pace.

Taking the Tough Road

Titan's early success is not surprising considering the hard work that characterized Gene Ray's life. Ray spent his early years in Murray, Kentucky, in a four-room farmhouse without electricity or running water. Finances were tight for the Rays, but there was never a question that Gene would attend college.

Ray did his undergraduate studies at Murray State University in western Kentucky. He majored in chemistry, physics, and math with the goal of becoming a chemical engineer. That goal shifted after a visit to the chemical engineering department at Vanderbilt University.

"When I saw how practical it was and how you needed to have skills with your hands, and [when I was faced with] big tanks of chemicals and vats

During his last year at Murray State University, Gene Ray (left) was named "Outstanding Senior Boy" and still calls the honor his proudest moment.

and things, it scared the hell out of me," said Ray. "I knew that wasn't what I wanted to do."[12]

Ray went back to Murray and took a hard look at which subjects he found the most challenging and interesting. He got straight As in math and chemistry but Bs in physics. Because physics was harder for him, he chose to pursue it.

That love for challenges followed Ray into graduate school, where he had to choose between experimental physics and theoretical physics. To pursue theoretical physics, students needed to find a professor to oversee their dissertations. There were only a few professors who were amenable, and they had to be talked into it. Once again, Ray chose to take the more demanding route.

He finished his dissertation in 1965, which was a good year to have a doctorate in physics. Many companies and institutions were looking for scientific whiz kids with Ph.D.'s. Ray was invited to 26 job interviews in all, and the number would have been greater had he had time for more. Half of the interviews were for teaching positions and half were in industry. Ray was more interested in industry because salaries were higher.

He accepted a position with the Aerospace Corporation, in San Bernardino, California, working in the area of nuclear weapon effects. It was a humbling start. Ray's boss gave him an unclassified book to read and asked him to plot some data. In his plotting, Ray misspelled a word. The boss was not happy. "He chewed me out," Ray said. "I'd never been so miserable in my life."[13] In the end, though, Ray found the job to be a positive learning experience and a stepping-stone to a greater challenge. He also found his boss to be a "great guy."

Meanwhile Ed Knauf, after graduating from the U.S. Military Academy (West Point) in 1965, joined the Air Force and began installing intelligence sites in the Far East. He did his graduate studies at MIT and was assigned to the Minuteman Program Office at Norton Air Force Base, where he was in charge of software development for Minuteman missile systems. It was at Norton that Knauf first met Gene Ray, who by that time was working as a civilian for the U.S. Air Force.

Ray had latched onto the government job after seeing a help-wanted flyer someone had left in the copying machine at Aerospace. As Ray told it, Secretary of Defense Robert McNamara had brought in a bunch of young Ph.D.'s who radically changed the planning process for the Department of Defense—a change that the services didn't like. The Air Force wanted to counter McNamara's "Whiz Kids," as they were known, so it began recruiting its own set of young Ph.D.'s. Ray qualified: He was young and had a Ph.D.

At age 29, Ray was selected to take charge of the Strategic Division of Operations Analysis and quickly found that he had gone from a humbling job to a heady one. Working for Ray were a half dozen people, including a lieutenant colonel, two majors, and two GS-15s. As civil service ratings go, Ray's PL313 was supposed to be equivalent to a two-star general. He even had an airplane at his disposal. One call and a helicopter would pick him up at the Pentagon and whisk him to Andrews Air Force Base, where a small military executive jet awaited.

Beyond the perks, Ray found the position to be "the best learning experience I ever had in my life. No one can understand how the Pentagon operates and how the military operates without being part of it."[14]

More than that, "It was pure challenge," Ray said. "I didn't know what couldn't be done. When you're a young, 29-year-old kid, you just don't have any idea what you can't do. So you accomplish a lot more than if you knew how difficult it was."[15]

While Ed Knauf was in the Air Force, he established a good working relationship with both

Years before they cofounded Titan, both Gene Ray and Ed Knauf worked for the Department of Defense. They were able to bring their knowledge of its inner workings to their work at Titan. *(Photo courtesy Department of Defense.)*

Ray and Col. Edgar Northrup, another of Titan's initial players.

Ray believed his greatest contribution at the Pentagon revolved around the strategic use of B-52s in Vietnam. The Air Force planned to use the aircraft to fly in low for certain targets. Using the best technology available, Ray determined that "flying high was without a doubt better than flying low."[16] The study took a year and a half and involved several hundred people and dozens of simulations. It was well performed and was well received by the highest ranks. In the end, the United States flew its B-52s high, and they were very effective and were not shot down.

Following that exercise, Ray might have been expected to make a career at the Pentagon, but he deliberately chose to leave. "When I went in, I was planning on staying anywhere from two to three years and absolutely no more than four," he said. "I didn't want to become a bureaucrat."[17]

Ray received several job offers, but one in particular riveted his attention. It came from J. Robert "Bob" Beyster, also a Ph.D. physicist, who had started a small scientific consulting firm called Science Applications Inc. (SAI, later SAIC) in San Diego in 1969. Beyster had previously worked at Los Alamos National Scientific Laboratory and Gulf General Atomic.

Rather than having a grand design, Beyster later observed, he started with a few contracts and a few people with ideas. "After a year, a surprising thing happened," he said. "We made a profit."[18]

Ray acknowledged Beyster as not only a brilliant scientist and executive but "the best one-on-one recruiter this world has ever known." Ray and Beyster had crossed paths a few times, and Ray saw that Beyster was bringing top talent into his new company. When Beyster came courting, Ray found it impossible to resist.[19]

What's more, Beyster was putting into practice the theory that people involved in a company should share in its success through stock ownership. That idea appealed to Ray. He would be SAIC employee number 32, which would put him in an excellent position should the company succeed. And it appeared that it would. SAIC had $250,000 in revenue in its first year and was halfway through a $1.2 million second year when Ray came on board.

Despite that bright outlook, Ray was nearly overwhelmed by culture shock during his first days at SAIC. Beyster had leased office space in the San Diego neighborhood of La Jolla with a terrific view of the Pacific Ocean—but not much else. The company's Web site tells the story.

The setting for our startup in 1969 was not much different than that of many companies that begin with a few dollars and a good idea. Small, austere offices with rented desks and typewriters that have seen better days. People furiously working days, nights and weekends, hoping to curb the anxiety over the future.

Ray had gone from having high-level Pentagon perks and a plane at his disposal to a tiny office with no secretary—and almost no furniture.

Ray called his study on the strategic flying of B-52s "by far the most productive" work he did at the Pentagon.

"I asked where the secretary was and was told they were downstairs on the first floor," Ray later recalled. "I said, 'Where's my secretary?' and was told those were the only secretaries there. I called down and said, 'This is Gene Ray. Would you come up?' and one said, 'What do you mean come up? I'm busy. Come down here.'

"So I went down, and my first question was, 'Well, how do I get a bookshelf?' 'Well, the way you get a bookshelf is you call Sears and Roebuck. They will deliver one that you put together yourself.'

"So I did, and I sat down on the floor the first day and put together this bookshelf. The second day, two of the shelves fell." Ray didn't bother fixing them for some time.[20]

When Ray joined the staff, SAIC was working solely on nuclear weapon effects, but Beyster wanted Ray to lead the company into a new area: systems. Ray was to bring in new business and run the business he brought in. If he didn't bring in new business, he wouldn't stay.

"I barely made it," Ray said. He joined the company in July 1970 and didn't bring in a contract until December 1970. "I brought in a contract on December 7. If I hadn't gotten it by December 31, I would have been out in all likelihood."[21]

Ray's first contract was from the U.S. Air Force and involved nuclear simulation. Once again, Ray and Ed Knauf made contact. "I was still at the Minuteman Program Office at Norton," Knauf recalled, "and Gene brought up a bunch of the guys from SAIC who were doing work for the Defense Nuclear Agency, which was important to what I was trying to get done." A few years later, in 1973, Knauf left the Air Force and began working for Ray at SAIC.[22]

After a few years at SAIC, Knauf met Albert Babbitt, vice president of technology for IBM. The government had recruited Babbitt to start a new governmental office called the Worldwide Military Command and Control System (WWMCCS, pronounced "wim-ics"), and Babbitt recruited Knauf as his deputy for the development of the architecture that would fit the military's various systems together. When Babbitt returned to IBM a few years later, Knauf went to work for San Diego–based Horizons Technology, where he started an office that worked mainly in command, control, and intelligence activities. Within a year, the office had grown to 25 people.[23]

Ray, meanwhile, had built his portion of SAIC's business to the point that he needed a staff of 500. It was much like constructing a business within a business. "You marketed. You managed. You brought in people," he said. "You ran the business, and that was good. It was a great education on how to do that."[24]

In fact it was such a great education that he realized he could run a company of his own. When Ray had been with SAIC 11 years, Beyster chose to segment the division Ray had built. Ray was not happy with the change and decided to leave the company.

He took a number of business lessons with him. Among them: "You found customers; you

One of Titan's first contracts involved analyzing target design and alternative basing modes for the Minuteman missile system. Pictured is an unarmed *LGM-30* Minuteman Intercontinental Ballistic Missile.

THE NAME GAME

Choosing a name for a company is never easy. Even keeping to basic guidelines—making it memorable, easy to pronounce, and meaningful to the industry—can be onerous.

Titan settled on its name more by fortuity than anything else, but it's a name that has become well respected in the industry—and over the years The Titan Corporation has grown to fill the colossal dimensions its name implies.

The first three or four names that were chosen had been taken. By the second week of May, the company had a half-dozen employees and a contract waiting for a company name.

Finally, Titan's employees were asked to come up with names. Cathy Pergam-Ball consulted a book on horse racing for ideas ("because horses have wonderful names," she said) and came upon the name Hyperion.[1]

Knauf wasn't sure what he thought of that suggestion, but he got on the phone with Ray, McDougall, and Ed Northrup to pass the idea along.

"How do you spell it?" asked Ray.

"I don't have the foggiest idea" was Knauf's reply.

"Well, then, that's a lousy one."

So they went back to the drawing board.

Meanwhile, Northrup had looked up Hyperion in the dictionary and discovered that it was one of the mythological Greek Titans.

"Why don't we call the company Titan?" he suggested.

Knauf's first reaction was "Oh, God, what an awful name. People will think we're old missiles or something."

But the founders submitted the name, along with five others, to a name search that would determine if any companies with those names already existed. Titan was the only name not in use.

"So we went with Titan Systems," Knauf said, "but initially Gene and I were sort of embarrassed."

They created a logo for Titan Systems and decided to call the company TSI, but TSI didn't stick. Within a few days, the founders had fallen into the habit of saying "Titan." Fortunately the name worked to the small company's benefit. After only a few days in business, Knauf remembered people in the government saying they'd heard of the company simply "because it was a name you ought to have heard of, right?"

At meetings between Titan and IBM, Al Babbitt, a vice president at IBM who promised to be one of the new company's biggest customers, jokingly introduced Titan to his fellow senior officers as "the small company with the precocious name."

But Titan's leaders didn't mind. "The name worked like a charm," said Knauf. "I told Babbitt we picked Titan because we didn't want to be just another three-letter company like IBM."[2]

understood what their problems were, what their issues were, and then you came in with a solution, and you delivered," Ray said. "That's still the way to do it. Hasn't changed." In addition, Ray learned about the value of teamwork. "On a major program, no one person does everything," he said. "It takes a team. That's probably my most important capability, to be able to pull together a team that can solve customers' problems."[25]

The Pace Is Set

Within its first few weeks of operation, Titan Systems accumulated a backlog of $559,000 in contracts. The largest contract, $414,000, had been its first, the one from GTE Sylvania for target design and alternative basing modes for the Minuteman and MX missile systems. It would last 12 months. Other contracts came from IBM, TRW, WGSA, and HTI/DNA. Even SAIC awarded the new company a small consulting contract, which marked an agreement between Ray and Beyster that Titan would not recruit from SAIC's ranks.

Business was so good that the partners began to feel somewhat overwhelmed. They called a meeting to discuss how to handle the workload. "We knew we had to sit down and come up with what the tasks were and get them assigned to people," Ray said. "Then, either Jack or Ed looks around and says, 'What the hell do you mean? We are the people.'"[26]

Vic Gogolak remembered well when he attained the position of vice president. "We were in business for about eight months," he said, "and one day Gene and Ed called me in to announce that I had become a vice president of the corporation. Then they explained that the criterion for the title was selling $1.5 million of business."[27]

Establishing such criteria was one of Titan Systems' first steps toward corporate maturity. In the meantime, however, the fledgling company had its share of growing pains. Linda Frady Keenan, who joined Titan in 1982 to fill in as an administrative person wherever she was needed, remembered how she acquired her first official position. "We forgot to put a fee on the first proposal we sent to IBM," she remembered. "IBM was nice enough to come back and say, 'We know you're a small company, but most of our contractors want some kind of profit after they've done a job. Why don't you add a fee to that?' So that was when Ed [Knauf] said, 'I think this is where you want to focus, Linda, because apparently we can't even get a proposal out the door to make money.'"[28] Keenan went on to become director of contracts for Titan Systems' Applied Technologies Group, where she is still doing an outstanding job.

Titan cofounders Jack McDougall (left), who had a background in nuclear engineering, and Ed Knauf (right), an astrophysicist, poured all of their experience and imagination into making Titan a success.

Despite such rookie mistakes, Titan's early contract strategy was aimed at keeping the cash flowing, something at which the company excelled.

"Everybody in the company had confidence in each other," recalled Ed Northrup, who became the fourth member of the Titan executive management group. Besides knowing Knauf and Ray through the Air Force, Northrup had worked with them at SAIC. "That confidence and the quality of our people enabled us to go out and get a lot of good business."[29]

Titan pursued subcontracts because they could be issued rapidly and generally paid promptly. Prime contracts could take months to approve, and payment was slow.

It was time for more hiring. Five and a half months after the company started (the end of the first stub period ending October 31, with the first full fiscal year beginning on November 1, 1981), the staff numbered about 24. Ray kept his agreement with Beyster that he would not recruit from Beyster's staff. Still, about 20 people from SAIC joined Titan in the company's first year. "People at SAIC knew we would not consider them for a job unless they wrote us a letter and specifically asked us if they could be considered for a job," Ray said. "Those were the only people we would interview."[30]

It was important for Titan Systems to be able to hire local talent. The cost of housing in the La Jolla area was high, and that was a roadblock to hiring from other areas of the country. Fortunately, there were educated and talented people in the area. About 28 percent of Titan's first employees had doctorates, and nearly half had more than 20 years' experience.

From the beginning, Ray, Knauf, and McDougall decided to share ownership of the corporation with employees.

"We were able to recruit very, very good people, share the ownership, and come up with a very strong staff right off the bat," Ray said. "That was our strength, and I think still is the strength of the corporation. You can see some of the employees we recruited early on and how well they did, not just at Titan, but at other places throughout their careers. I think it is indicative of the level of people that we were able to recruit."[31]

Titan was able to retain such gifted employees by fostering a cooperative rather than a competitive environment. "The model that Gene and Jack and Ed were familiar with from SAIC was very internally competitive," Pergam-Ball said. "And we all decided that was not healthy, that we ought to be focusing our competitive juices on the outside and not on each other. To foster that, we ended up doubling the sales credit and then splitting it. So if I had a target, and I wanted Richard Llewellyn's help, we could split the sales credit—each getting 100 percent—and did not have to compete with each other. He was incentivized to give me people if I needed people. It had a very positive effect."[32]

As for management of the company, Ray, Knauf, and McDougall divided responsibilities based on past experience. Knauf, as executive vice president, chief operating officer, and chairman, was largely responsible for bringing in the contracts. McDougall was executive vice president and secretary and managed the day-to-day activities of the business. Ray served as company president and chief executive officer. The three men developed four basic operating principles for the company:

Excellence in all endeavors
Customer satisfaction
Corporate integrity
Respect for the individual

"We took those principles very seriously," Ray said, "and they're still the operating principles for the corporation."[33]

The long-term plan was to provide steady growth at an average rate of between 40 and 50 percent per year with a corporate structure that would allow for an eventual initial public offering (IPO) or acquisition by a reputable firm. The 10-year goal was $22 million in revenues.

By November 1, 1981, the beginning of its first full fiscal year, Titan had seven contracts and $793,000 in revenues.

Titan owed much of its early success to the high caliber of its professional staff. In 1984, 21 percent held Ph.D.'s and 50 percent had more than 20 years' experience in their chosen fields.

CHAPTER TWO
THE EARLY YEARS
1982–1984

We performed, and we performed extremely well.... We got the job done for the customer, on cost, and on schedule.

—Gene Ray, 2001, speaking of Titan's early years

THE FIRST FEW YEARS OF BUSIness marked a heady time for Titan Systems. Solid contacts in the defense industry scored the initial contracts, and when those contracts were executed effectively, more jobs rolled in.

Company founders Gene Ray, Ed Knauf, and Jack McDougall worked long and hard in the early days. "We were as highly motivated as you can ever be," said Ray, remembering the inaugural years. "Adrenaline flowed like crazy during that time."[1]

Titan staff members were equally dedicated and involved in the company's adventurous culture. "We would take on jobs that DoD [Department of Defense] quite frankly didn't think could be done but had to try anyhow," said David Porreca, a former major with NATO who became the company's chief scientist in 1982. He later became president and CEO of Titan's Cayenta subsidiary. "It was very experimental," Porreca continued. "But it was very scientific. There was a lot of basic as well as applied technology involved in the work we did, and it was truly some of the most fun any of us have ever had in our lives—not just because of the culture but also because the government was attempting to do things it hadn't done before."[2]

"We couldn't keep up with the opportunities," said retired Vice President Karl Gould, who, as Titan's 14th employee, set up an office in Colorado Springs that grew to employ about 50. "It was absolutely great. Reagan had decided to win the Cold War and spend the money to do it. We had tons of talent, and we didn't have a lot of bureaucracy. You didn't sleep a lot, but it was probably the richest professional time that I've ever had."

Gould went on to discuss Titan's patriotic role during the Cold War.

I'm a nuclear warrior, a guy who's spent most of his professional life trying to win the Cold War, and, disproportionate to its size, Titan contributed to that in ways that will never really be talked about. Titan helped the government find ways to make the command and control infrastructure of the United States more survivable, less vulnerable. We contributed to that, and that's something that endures forever.[3]

Ray was able to attract enthusiastic workers like Porreca and Gould through skillful recruiting and the prospect of ample rewards. "Recruiting was where I spent a lot of my time early on," he said. "That's always the key—getting the right people in."[4]

Gene W. Ray, pictured here in 1984, has led Titan as president and chief executive officer since the company's founding.

Also, the plan from the beginning was for Titan Systems to go public. Employees who helped make Titan a success could expect to reap financial benefits.

"We were creating very real value," Ray said. "We told [prospective employees] that we were going to build a major corporation. Early on, we set a goal to get to $1 billion in revenue."[5]

The first step in reaching that goal was to define the space Titan Systems would occupy in the business world. The founders had concluded that the first generation in the high-technology and systems engineering industry was coming to a close. They saw that Titan could become a leader of the next generation, focusing primarily on solving complex military and civilian problems of information synthesis, global communications, intelligence, and technology development.[6]

From the beginning, Titan Systems took risks. "People didn't know how to take different software semantics and translate them into a higher-level language," said Porreca. "And in those endeavors, we went in, quite bluntly, and said, 'We're not sure we know how to do it either, but we think most of the current approaches are wrong.' So we were willing to gamble and try something new."[7]

The company's approach would be soft, not hard. Titan was to provide "systems engineering, systems integration, and systems analysis, and that's not a hard product taken off a shelf," Ray told the *San Diego Union* in 1983. "It's a process that one has to go through."[8]

High-level software development would be part of Titan's plan, for as Jack McDougall explained, "We are no longer an industrial-based society but have become an information-based society, indicating the growth of the relative importance of software."[9]

Mary Kearney, one of Titan's employees who had also worked at SAIC, wrote about the company's early years in a 1985 *San Diego Corporate Guide to Business*:

> *TITAN was founded in 1981 in the belief that a unique window of opportunity exists for high-technology systems development and integration in government defense-related programs. The aerospace industry has just completed its first generation, having begun some 25 years ago with the successful initiation of our nation's space programs. Many technical and management professionals, such as the founders of TITAN, literally grew up in that new industry.*[10]

Titan Systems was not to be confused with a consulting company or think tank. It would provide a service. As expressed simply in the 1984 Titan annual report, "TITAN was founded to take advantage of the emerging technologies of the Information Age by providing sophisticated system engineering for the Department of Defense and other government agencies."[11]

C³I

The company's primary customer in the early years was the military's growing command, control, communications, and intelligence community, known as C³I. Experts predicted this marketplace would grow 35 percent over five years, from $8.2 billion in 1983 to $11 billion in 1988. Intelligence programs would be the main beneficiaries of increased C³I spending.[12]

The same group of market analysts expected the United States to spend nearly $32 billion by 1988 on aircraft, satellites, and systems for reconnaissance and surveillance. The military

Much of Titan's early work involved command, control, communications, and intelligence (C³I) for the U.S. military. In 1984, the company designed and integrated a prototype secure full-motion video-teleconferencing system that analyzed crisis situations and supported operational military planning.

CHAPTER TWO: THE EARLY YEARS

intrusion detection market was expected to grow by 34 percent.[13]

These military developments were being pushed by the administration of President Ronald Reagan to battle the spread of Communism throughout the world. The country embarked on a huge buildup of both conventional and nuclear forces. In 1983, Reagan initiated the much publicized Strategic Defense Initiative (SDI), or "Star Wars" missile defense program.

"It was the height of the Cold War," Porreca remembered. "There was a serious coterie of people in the country who really believed that time and resources were on the side of the Soviet Union. We knew we had to think smarter or else the Soviets could always outproduce us when it came to tanks and planes and everything else. We had to build better things, smarter things."[14]

All of these factors indicated that Titan Systems was building a business for a solid market. "The complexity of a modern defense program demands sophisticated information systems based on computer technology to assist military and civilian commanders in managing and controlling weapons (aircraft, ships, tanks, and missiles) and tactical forces," Kearney wrote in the *Corporate Guide*. "The C³I systems are regularly upgraded to enhance safety and security, as well as to increase effective-

Above: Titan supported development of alternative Strategic Defense Initiative (SDI) plans by developing analytical tools and databases that simulated missiles and spacecraft.

Below: Titan specialized in systems engineering for advanced naval vessels as well as for tactical aircraft and C³I systems.

ness and accommodate new or advanced weapon developments. TITAN emphasizes communication system planning and integration plus software design and development. It also uses advanced decision aids incorporating artificial intelligence techniques."[15]

Titan's leaders planned a wide business footprint for their company and divided it into three segments: advanced technology (20 percent); system development and integration (60 percent); and system analysis and specialized products (20 percent).[16]

Advanced Technology

Titan's founders were adamant that the company become a world technology leader by focusing on emerging technologies that were likely to become prevalent. By the end of 1984, Titan's advanced technology segment had singled out three areas: electro-optics, artificial intelligence (AI), and directed energy.

Titan contributed to the development and testing of ground-based, air-based, and space-based electro-optics systems and the application of advanced optical sensors to strategic and tactical defense. The company also supported research and development of laser and communications systems. Major initiatives included signal processing and advanced sensor technology.[17]

In addition, Titan developed software and computational resources to support AI environments. Those environments included interpretive languages, rapid prototyping capabilities, semiautomatic knowledge acquisition, and research/engineering tools for expert system building. Titan's Adaptive Maneuvering Logic training system simulated an intelligently interactive opponent in a flight simulation. The system was so effective (the simulator always beat the person) that the simulator's capability had to be degraded in order to train pilots. TACTICIAN, another Titan system, integrated several AI concepts into a combat-decision aid and was capable of testing alternate solutions using a battle simulation. Titan's INTERPRETER system, which increased the understanding of message content, permitted an analyst to determine specific information on a narrow subject extracted from large amounts of data.[18]

Titan also sought to improve weapon concepts, particularly for the Star Wars initiative, and had its own scientists working on directed-energy concepts. In 1983 the company augmented its capabilities by acquiring a complementary hardware product company, Beta Development Corporation, located in San Francisco's Bay Area.

Beta Development was a leading producer of specialized system components for high-power laser and microwave weapon systems, particle beam accelerators, and fusion research devices, including high-power electron guns, special power conditioning and triggering systems, and a variety of laser pre-ionizers.[19]

Beta Development had been founded in 1980 by an entrepreneur named Gary K. Loda. According to Loda, the company's long-term goal was to irradiate food. "But we knew that goal was going to take a long time," said Loda, "so in the interim, we did Defense Department–related activities," which it was doing when Titan Systems acquired it. In 1984, Mike Dowe, vice president for corporate development, Gary Loda, and an outside consultant put together a business plan for building a food irradiation plant

and presented it to Ray. It was decided that the plan was too early, but the seed had been planted. Eventually, the electronic beam technology developed by Beta would lead to Titan's patented SureBeam® system that killed harmful microorganisms in food, and Loda was there to see it happen. He would stay with Titan as president of the Beta subsidiary until 1990, retire for seven years, and come back in 1997 to help launch SureBeam.[20]

System Development and Integration

Often the government's system development and integration work required specialized software, and Titan was able to provide that specialized assistance. Demand was great enough, in fact, that by 1985 system development and integration would comprise more than 75 percent of Titan's business.

The company had its own software developers and added to its software and systems simulation expertise with the acquisition in 1983 of Compunet, a recognized leader in strategic defense systems software.

One Titan project for the U.S. Army evaluated the test software for high-speed integrated circuit chips and recommended software with side applicability, easily convertible to the proposed standard Defense Department software language. Titan also developed a software simulation program for the Army's Short Range Air Defense System that modeled all parts of the system and then identified and evaluated communication flow and data processing needs for each part.[21]

In addition, Titan provided system development and integration for a number of Department of Defense weapon systems, such as the Army's advanced ballistic missile defense systems. Titan evaluated command and control systems and performed life-cycle cost analyses of system concepts.[22]

Projects for the Air Force included the development of strategic operational concepts, threat modeling, survivability evaluation, and determination of logistics requirements.[23]

Titan also worked with a nonmilitary client in the system development and integration area. Under a contract with the nuclear power generating station at San Onofre, California, the company operated a computer center, developed analysis programs, and established and maintained a large database monitoring the environmental impact of the station on the ocean.[24]

System Analysis

Titan Systems undertook a wide variety of system analysis projects, focusing on issues related to implementing future policies and system acquisition strategies. Typically those issues involved identification and assessment of operational requirements, technical alternatives, life-cycle cost, and threat and policy implications.[25]

Titan performed threat and survivability analysis for advanced systems such as the Hard Mobile Launcher and the Small ICBM Program. The company also supported agencies involved in assessing the implications of alternate U.S. arms control options.[26] In addition, Titan Systems supported a variety of studies of future C^3I and weapon system alternatives such as the development of alternative SDI architectures. The purpose of these efforts was to define investment alternatives, including cost/benefit evaluations and operational concepts.[27]

This high-power electron gun is just one example of Beta Development's specialized hardware. Beta was one of Titan's first acquisitions. It produced system components required for high-power, directed-energy weapon systems and fusion research devices.

Titan's C³I systems assisted the military in managing and controlling missiles, such as this Tomahawk cruise missile, as well as aircraft, ships, tanks, and tactical forces.

Titan developed several analytical tools and databases to support system analysis projects. Those included simulations of missiles and spacecraft, models for system survivability production, and an evaluation technique for assessing the relative worth of alternative systems.[28]

Specialized Products

Always looking to the future, Titan Systems worked with its wholly owned subsidiary, Beta Development Corporation, to expand its specialized products capabilities significantly beyond accelerator and laser components and power systems.[29] Titan's leaders planned for their special products to lead them into both industry and government markets.

While Beta was developing three excimer laser systems for the U.S. Air Force, it was also implementing a process for repairing defects in the lithographic masks used by manufacturers of VLSI (very large scale integrated) circuits and related devices. Its Series 200 Mask Repair Station was available to both government and industry.[30] Beta also developed power systems and power-system components for high-power laser and particle accelerator devices,[31] the same technology that would lead to SureBeam.

Another of Titan's specialized products was a defense-related medical product. Under contract with the U.S. Army Medical Research Center, Titan delivered a prototype noncontact heart-rate monitor. This physiological sensor could measure human vital signs without actual body contact. The device could be used by the military for monitoring people wearing protective clothing. But it also had potential nonmilitary use. The San Diego Police Department was testing it for use as a lie detector. Titan was also exploring the use of the device in medical settings.[32]

In addition, Titan designed and developed computer interface hardware and software that allowed mainframe STRATUS computers to interface with alternative communication media.

Company Success, Company Culture

Titan's strategies paid off in a big way in the early years. The company's achievements significantly exceeded its growth and performance goals. Contract revenue jumped from $5.4 million in 1982 (its first full year of business) to $13.5 million in 1983 and to $24.2 million in 1984. Contract backlog at the end of 1983 stood at $14 million. For the 1984 fiscal year, Titan's revenue increased by 80 percent, and contract backlog was at $30 million.[33]

The number of Titan employees grew from 119 in 1982 to 211 in 1983 and to more than 300 in 1984. Realizing that every company has its own personality, Titan's founders aimed at creating a corporate culture that would encourage managers and employees to keep up the good work. They recruited top talent, and, once on board, employees were encouraged to follow Titan's four operating principles, which new employees received. Ray expanded on the operating principles in the 1984 annual report:

Excellence in all endeavors. The Titan name must be synonymous with superior performance in every area of endeavor.

Customer satisfaction. Every customer is entitled to a Titan commitment to provide quality products and responsive service.

Corporate integrity. Titan is committed to the highest standards of professional integrity in all facets of its business activities.

Respect for the individual. Every individual must be treated with respect and be provided an equal opportunity to succeed.[34]

To reward employees for the company's success, Titan significantly enhanced its benefits package in 1984. The company's contributions to both the profit-sharing and employee stock option plans were double what they had been in 1983. Managers were given a new stock options incentive program, which offered rewards for obtaining new or follow-up business. Titan also expanded its bonus program, which allowed management to award on-the-spot bonuses to employees for outstanding accomplishments. The company even loaned a total of $220,000 interest free to employees for them to buy personal computers.

Encouraged by the company's early success, Titan's leaders recommitted to their strategy of management by objective. They used a systematic, goal-oriented approach to develop Titan's operating plan and budget for each fiscal year and to develop a long-term strategic plan for corporate growth. A number of senior employees were involved in this planning process, and each committed to achieving the company goals.

As was his custom, Ray showed his appreciation of Titan's employees through not only generous rewards but generous words as well. "I would like to express my appreciation to the entire Titan team for their commitment and hard work," he wrote in his 1984 President's Letter. "By cooperation and dedication to a high standard of excellence, we have made significant contributions to national security and established a solid base for future growth."[35]

The San Diego Scene

Titan's early success reflected what was transpiring on the San Diego business scene in the early 1980s. San Diego was still predominantly a defense town, but defense was no longer the only game in town. Aviation and tourism had become substantial industries. The city, like the rest of the country, had rebounded from a recession, and the economic future looked limitless. San Diego leaders still fought the wildfire growth that had overcome other areas, such as Los Angeles and Orange County, but some growth seemed inevitable.

Titan had two major centers, one in San Diego (top) and the other in Vienna, Virginia, near Washington, D.C. (bottom).

Major corporations eyed the area as a potential location. Housing prices were relatively high, but the climate and quality of life drew many. A new convention center was in the works, a development that was good news for San Diego's business community.

For its part, Titan located its headquarters in San Diego's Golden Triangle, an area close to the University of California at San Diego that was beginning to draw top technological firms and talent. From Titan's tall, prosperous-looking headquarters on Towne Centre Drive, the future looked very bright indeed.

By the end of 1986, Titan had its strong management team firmly in place. Leading the company were Sid Webb (left), chairman of the board, and Gene Ray (right), president and CEO.

CHAPTER THREE

MERGING INTO THE FAST LANE

1985–1986

We are now more integrated and streamlined and have reached the critical mass where specific large programs will be targeted and pursued.

—Gene Ray, 1986

BY EARLY 1985, Titan Systems was on the fast track to becoming a major corporation, a business entity that would have the capabilities and the financial base to bid on larger defense projects.

Fittingly, the year began with a victory—a major Star Wars contract that teamed Titan with Rockwell International and Lockheed Corporation. In the words of Gene Ray, the companies would "lay out the architectural designs for what a defense system might look like for the future."[1]

It was a step forward, but Titan's leaders dreamed of even greater growth, and that would demand more capital than the privately held corporation could muster. They felt that the time was right to steer Titan toward an initial public offering (IPO). Some influential people supported the idea, among them former Secretary of Defense Harold Brown and future Secretary of Defense Bill Perry.[2]

Moving toward that goal, Titan's leaders met with financial analyst Andy Gaspar from Warburg Pincus, at the time one of the largest venture capital firms in the United States. The meeting went well but underscored the learning curve that Titan's leaders faced. "In the first briefing, we tried to say that we wanted liquidity for our shareholders," Ray recalled. "But instead of calling it liquidity, we called it fluidity, and we didn't know the difference."[3]

Titan's management team quickly learned the appropriate lingo and went on to choose underwriters for the IPO. Kidder Peabody was set to be the lead underwriter, and Robertson Coleman was to be the colead.

A New Path

In the midst of IPO planning, Ray received a phone call that would change Titan Systems' course. On the other end of the line was Rolf Erikson, who handled mergers and acquisitions for a company called Electronic Memories and Magnetics (EMM), in Encino, California. EMM was a 16-year-old manufacturer of computer and memory products, mainly for the DoD, and it was about 3.5 times the size of Titan. It had been successful but had recently fallen into inertia. The founder and chairman had retired, and J. Sidney "Sid" Webb had taken his place. (Webb had retired in 1981 as vice chairman of TRW after building a very successful billion-dollar commercial technology business.) After seeing Titan's 1983 annual report ("our first classy one," according to Knauf), Erikson and Webb wanted to talk business. Erikson thought he

---

After Titan merged with Electronic Memories and Magnetics (EMM) in 1985, the name of the company changed from Titan Systems to The Titan Corporation.

could convince Ray to merge with EMM by inviting Ray to be the company's CEO.

"I tried to talk Gene out of going public and into coming to join EMM, which was on the New York Stock Exchange," Erikson remembered. "He expressed some interest, but not that much. Then I talked him into seeing our headquarters, and he met with Sid Webb."[4]

Webb and Ray were not strangers. They had been introduced the year before by an investment banker and had shared lunch. "We had a nice lunch; I guess we liked each other," Ray said of the earlier meeting. "But we didn't see how to put anything together. I forgot about it. He forgot about it."[5]

The two had more to talk about by March 1985. Both men saw the potential for a fruitful marriage. Their companies might do together what they were unable to do separately.

After rapid due diligence on both sides, Ray held a meeting to brief Titan's top people on EMM and to discuss whether Titan should merge with it. "I'll never forget that meeting," said Jack McDougall. "Karl Gould was very concerned that the two companies wouldn't work well together. He thought if we merged, we would have all these problems."[6]

The meeting started at 9 A.M., and McDougall watched Gould closely during the briefing. "He was getting more and more interested in the merger," McDougall said, "and at noon he had to leave to catch a plane. So just as he's about to walk out the door, Gene turned to him and asked him what his thoughts were. Karl gets kind of a grin on his face and says, 'The urge to merge is beginning to surge.'"[7]

While not exactly great poetry, Gould's summary of the meeting was indicative of how enthusiastic Titan's people felt about the prospect of joining with EMM. Shortly thereafter, the companies agreed to a reverse merger, in which Titan Systems, the smaller company, would absorb EMM. The merger was finalized two months later, in May 1985. Because EMM was already publicly traded, Titan became a public company without holding an IPO. Titan shareholders received 2.1 shares of EMM common stock for each Titan share. Following the merger announcement, EMM's stock jumped $2\frac{1}{2}$ points.[8]

SureBeam board member John Arme, who as a partner with Arthur Andersen and Company served as Titan's independent auditor, noted that probably less than 5 percent of the companies going public in the mid-1980s did so through a reverse merger. "Sometimes in the investment community it was viewed as the wrong way to go public," he said, "because basically you are not introducing your company to the public market to let the investors determine whether they will accept it or reject it and set the price." But, Arme pointed out, Titan's proven success quickly overcame any queasiness on Wall Street.[9]

Ray remembered how difficult it was to call the investment banks to tell them Titan was canceling its IPO road show days before it was scheduled to begin. To keep good relations, Titan paid the bankers' out-of-pocket expenses related to the IPO process.[10]

The newly merged company was named The Titan Corporation—or TTN on the New York

---

From left: Sid Webb, chairman of the board; Ed Knauf, executive vice president; Rolf Erikson, senior vice president, corporate development; Gene Ray, president and CEO; and Jack McDougall, president, Titan Systems

CHAPTER THREE: MERGING INTO THE FAST LANE

Titan developed advanced software for military computers, communications, navigation, radar, and signal processing. Its software could be found in the B-1B, which is an important part of the United States' long-range bomber force.

Stock Exchange. Titan Systems became a subsidiary of The Titan Corporation, and EMM's various companies became Titan divisions. With a combined $112 million in sales and a work force of 2,200, the new company emerged as one of the largest in San Diego County.

"What this merger really does is to open a whole sector of business to the combined company that wasn't open to either one before this," Ray told the *San Diego Union*. "The merger opens the defense sector to EMM's products that they had not marketed to the defense community before, and it opens some commercial markets for us that we couldn't have looked at."[11]

Webb was named chairman of the new company, and Ray became president and CEO. Headquarters were temporarily located in Encino before moving to San Diego.

Titan saw several immediate benefits from the merger. EMM had $21 million in cash that would allow Titan to grow into a larger, stronger business.[12] ("That was more cash than we could have raised in an IPO," Ray noted.[13]) Furthermore, EMM's SESCO (Severe Environment Systems Company) division, which manufactured computer boards using Intel technology, added military electronics manufacturing operations to Titan's portfolio. What was more, Webb, a former vice chairman of TRW, brought along a razor-sharp business sense. He would put it to good use at Titan.

Sid Webb

Oddly enough, Sid Webb was supposed to be retired. Intent on relaxing and cataloging his

Sid Webb at Titan's 1986 holiday party. Titan's scrapbook describes this photo as the first Titan Corporation Christmas party at the "posh chez Ray."

collection of 4,500 bottles of fine wine, he had stepped down as vice chairman of TRW in 1981.[14] During his tenure there, he had bought nearly 20 companies, sold eight more, and helped build the company into a $5.5-billion-a-year defense and electronics giant.[15] After 27 years with TRW, he had been in line to become CEO, but taking that position would have required a move to Cleveland—a change he didn't want to make.

Wine cataloging satisfied a savvy business pro like Webb for only so long. He quickly grew bored and in 1984 was appointed chairman of EMM. It was just the kind of challenge he needed to get the adrenaline flowing again. "I saw a company that was stagnant," he said. "There was no cohesion, no strategy . . . only a hodgepodge of different businesses."[16]

EMM had been a highflier in the 1960s, but it fell to earth with a thud before the 1970s had ended. Its magnets and electric motors had become commodities, and its computer memories grew obsolete. The company's annual sales had peaked at $131 million in 1978 but by 1983 had dropped to $79 million.[17] Revenue had risen to $96 million by the time Titan and EMM merged.

Webb believed that one way to rejuvenate EMM would be to lead it into the $10 billion market for military communications systems and Star Wars weapons.[18] A second way would be to link with Titan. He was right on both counts.

"I think most people had kind of accepted the idea that EMM was a lackluster, non-growth company," Webb told *Barron's* magazine less than a year after the merger with Titan. "They're surprised how much we've been able to turn it around in just a few months."[19]

Ray was excited to join forces with a business force like Webb. "He was a terrific guy, an absolutely fantastic individual," Ray said. "He was a tremendous help to The Titan Corp., and to me through the years."[20]

Behind the Scenes

There were a lot of pluses to the Titan-EMM merger, but behind the rosy surface were a few negatives as well. The day the merger was finalized, May 29, 1985, was filled mainly with congratulatory phone calls, but Robert Hanisee, one of the few business analysts who followed both companies, asked Ray, "Why in the hell did you do that? Why did you tie yourselves up like that?"

"That gave me pause for about 30 seconds," Ray said. "But it was too late. The deal was done."[21]

As it turned out, Hanisee's fears that the merger would bring Titan down were not realized. As proof positive of his confidence in Titan, Hanisee would join the company's board of directors in 1998.

Still, in the three months following the merger, Ray got a clearer picture of EMM's strengths and foibles and realized that most of its businesses would have to be divested. "They were in eight different industries," Ray recalled. "It was obvious that some of those businesses had to be sold. Our plan was to sell off the old technologies and keep one that really was in the area we were looking for, which was building military computers [SESCO]."[22] Eventually, Titan sold that one too.

The Analysts

Though there were some dissenters, most business analysts gave the Titan-EMM merger a thumbs up. *Value Line* analyst Lucien Virgile compared the new Titan to a young TRW, saying, "Titan may prove to be the kind of stock to be held for many, many years. . . . By 1988–90, the price

---

Since its earliest years, Titan's board of directors has comprised some of the country's most brilliant minds. In the mid-1980s, the board sometimes held rooftop luncheon meetings. Clockwise from lower left: Gene Ray; Daniel Fink, former senior vice president of corporate planning and development, General Electric; Harry Derbyshire, former executive vice president of Whittaker Corporation; Joseph Caligiuri, then executive vice president of Litton Industries; and James Woolsey, a partner in Shea and Gardner and later a CIA director.

could well double, but, before selling . . . take a fresh look at where the company is heading."[23]

"Titan may live up to its name," Virgile continued. "In the past, a number of companies have used their mostly military technologies (aircraft, radar, communications, etc.) as a base for becoming military-industrial giants. We think that Titan has similar goals based on an important current state-of-the-art technology—sophisticated system engineering."[24]

Dillon Read and Company was also enthusiastic. "Titan brought to the merger a strong management team, a record of rapid growth in sales and earnings, and the need and ability to make successful acquisitions," Dillon wrote. "EMM contributed cash, a listing on the New York Stock Exchange, manufacturing facilities in the Far East, divestiture possibilities that could create additional cash and tax-loss carryforwards, and marginally profitable businesses with the potential for improvement."[25]

Analyst James J. Horan of Dillon gave Titan a "buy" recommendation and said that Titan's combined management team "demonstrated the ability to implement a high-growth strategy."[26]

Titan was not alone among San Diego companies regrouping in 1985. A number of area businesses sought safety and growth by joining with other companies or diversifying their businesses. The area's largest publicly traded corporation, Signal Companies, of La Jolla, with $6 billion in revenue, merged with Allied, of Morris Township, New Jersey, to become Allied Signal. Rohr Industries took steps to become not just a subcontractor for aircraft parts but a systems supplier. Cipher Data Products laid off 13 percent of its workforce to keep revenue growing.[27]

Trolling for Talent

With the merger complete, Ray and Webb began a search for the talented executives and scientists who could help turn their dreams into reality.[28] A key addition to the Titan board of directors was TRW cofounder Simon Ramo (the "R" in TRW), Ph.D., who was elected in December

CHAPTER THREE: MERGING INTO THE FAST LANE

1985. Dr. Ramo had been a pioneer in electron microscope development and served as a top government scientist during the early intercontinental ballistic missile programs.

"The addition of Dr. Ramo to our board is one more major move in helping Titan achieve its strategic objectives," Webb said in a company press release. "Dr. Ramo is a world-famous scientist, engineer, business entrepreneur, and author. His brilliant career spans General Electric, Hughes Aircraft, TRW, and Bunker Ramo."

A second key Titan addition was Gerold Yonas, Ph.D., former chief scientist of the Pentagon's Strategic Defense Initiative organization and a top civilian Star Wars scientist, who joined the company in August 1986. Yonas had not only held top Star Wars positions but spent five years at Physics International directing research in high-current electron beam and radiation effects simulation. Yonas was named a Titan vice president.[29]

"I view Titan as an opportunity to apply SDI technologies beyond military aspects," Yonas told the *New York Times*. "Titan offered me an opportunity to apply my technical expertise to a broader area than I have in the past."[30]

Analysts approved of Yonas's appointment. "He is certainly a feather in their cap and he'll probably

---

Opposite: Titan supplied core memories for the Programmable Display Generator, part of the avionics for the U.S. Air Force's F-16 fighter.

Below: The Titan Corporation commissioned this drawing from artist Loretta Dovell as part of its five-year anniversary celebration in 1986. The road on which the caricature of Gene Ray stands represents Titan's progress since its humble start in Ed Knauf's home in Vienna, Virginia. The road is paved with Titan's various acquisitions.

Titan's Famous Humor

Left: Gene Ray (right foreground) and other Titan employees celebrate the merger with EMM. Titan's people have always worked hard and played hard and have not been opposed to some lighthearted behavior.

Below: Mary Kearney, director of personnel, was one of Titan's earliest employees and was known company-wide for her patience, guidance, and—of course—her sense of humor. She died in a tragic plane crash on September 13, 2001, in Mexico.

WHEN DESCRIBING THE KIND OF people who make up a technology company like Titan, the tendency is to think of a group of stuffy intellectuals. But that is hardly the case at Titan. Though the men and women who make up the company are certainly brilliant, they are anything but dull.

"We were notorious for working hard and playing hard," said Nancy Jenkins, who was hired in 1985 as Ed Knauf's assistant. "We would work seven days a week and half the night on proposals and contracts. But we had a great time."[1]

Humor was and still is a key ingredient in Titan's culture. Chairman Sid Webb had a penchant for telling jokes, and it wasn't uncommon for management meetings to wrap up as comedy sessions.

Twenty years after the event, Titan veterans still chuckle about Mary Kearney's hiring experience. Ray, Knauf, and McDougall were having dinner and cocktails at the Knauf home one evening and decided they wanted to hire Kearney, who had worked for Ray and McDougall at SAIC. It was after midnight, but they called her at home to offer her the job. The next morning, each man realized that perhaps they'd had too much to drink the night before and that Kearney hadn't taken them seriously. Each of them called her independently to tell her that, yes, they really did mean to offer her a job. At Titan's 20-year reunion, Kearney told with affectionate humor how, after receiving the three phone calls the next morning, all to tell her the same thing, she had wondered what she was getting into. Kearney went on to lead the human resources department and became the lead administrator of Titan's operations on the West Coast.

Lawyers, as a rule, are not known for their sense of humor, but the men and women who made up Titan's legal department were hardly lacking in comedic wit. When an Australian company wanted to use Titan Beta's accelerator technology to irradiate the back ends of sheep to keep wool from growing where the sheep's defecation was sticking and causing infection, Titan's legal department under Marsh Nelson couldn't resist adding a touch of humor to the rather awkward case. Nelson dubbed it Project ROBOS, which stood for Radiation of the Butts of Sheep. The case never got past the legal department, but it did lead to quite a few chuckles.[2]

CHAPTER THREE: MERGING INTO THE FAST LANE

From left: Simon Ramo, cofounder of TRW, joined Titan's board in 1985. James Woolsey, undersecretary to the Navy during the Carter administration, became a board member in 1982. Ted Kavanaugh, former deputy controller at Allied Signal, joined Titan in 1986 as senior vice president and chief financial officer. Gerold Yonas, former chief scientist of the DoD's Strategic Defense Initiative organization, joined Titan in 1986 as a vice president.

start helping them straight off the bat," said analyst Todd Kelly with Conning International.[31]

There were to be many more feathers in Titan's cap. Ray and Webb sought highly educated, elite candidates from business and the military—and they found them. R. James Woolsey, undersecretary to the Navy during the Carter administration, had been a Titan board member since 1982. Gerry Keahl, former manager of the Scientific Instruments division of Beckman Instruments, became senior vice president of operations. Gene Gilliland, a former General Electric executive, was named senior vice president and general manager of Titan's Washington, D.C., office. Edward F. "Ted" Kavanaugh, former deputy controller at Allied Signal, became senior vice president and chief financial officer. D. Marshall Nelson, former vice president and general counsel for SAIC, became Titan's vice president, general counsel, and secretary. Albert Babbitt, former vice president of IBM's Federal Systems Division, who had worked with Titan since its earliest days, became special assistant to the president.

Bill Zettinger, who later became vice president of programs for Titan's Linkabit subsidiary, began consulting for Titan in 1985 and remembered what an easy decision it was for him to come on full time. "Titan fosters an environment that promotes innovation and flexibility," he said. "It was very refreshing after coming out of a big organization and big culture. At Titan we could do what we needed to do, and though in the early days we didn't have a lot of resources, somehow we knew we were going to do it—and we did."[32]

Adding and Subtracting

In mid-1985, Titan's leaders established four corporate goals: stimulate internal growth, maintain profitability, make strategic acquisitions, and divest operations that did not fit the company's growth plan.

Titan's goal was to reach $300 million in annual revenue.[33] To achieve this, Titan would need to make more acquisitions and ratchet up performance. In late 1985 and 1986, company leaders embarked on an acquisition program, purchasing six companies.

Then-CFO Ted Kavanaugh remembered the intensity with which Titan's leaders evaluated potential acquisitions. "We [senior management] met about once a week to get an update and to go over a list of anywhere from five to 10 acquisition candidates," Kavanaugh said. "Some were initial contacts, and some we were trying to close. We were looking at those companies that would fit with the company's long-term strategy."[34]

Or, in the words of Gene Ray, "We looked for strategic fits. The first thing you do is decide where you want to go and what your strategy is for getting there and then look for companies that fit that strategic picture."[35] Titan wanted to be a leader in defense communications and technology, so it went after companies that would help in that effort. In short order, Titan acquired a number of defense communications and technology companies.

Aeronautical Research Associates of Princeton (ARAP) was a high-technology engineering research firm with offices in Princeton, New Jersey, and McLean, Virginia. Ninety percent of ARAP's work was for the DoD.

Defense Systems Corporation (DSC), of San Diego, was a national leader in the application of

Engineer Julie Martin's work in electro-optics included testing an environmental protection system for a unique electro-optical sensor.

artificial intelligence to avionics. The company was a key team member of the Pilots Associates Program under development for the Defense Advanced Research Projects Agency (DARPA) and the U.S. Air Force's Wright Aeronautical Laboratories (AFWAL).

Spectron Development Laboratories, of Costa Mesa, California, was a leader in the application of laser technology to experimental aerodynamics, fluid mechanics, and electro-optics and for flow visualization, particle field diagnostics, and optical inspection. This acquisition was meant to enhance Titan's capabilities in the areas of electro-optics and laser technology for SDI, C^3I, and instrumentation applications.

California Research and Technology (CRT) was a national leader in the development and application of large computer codes for the modeling of scientific and engineering problems in the areas of explosive processes, nuclear effects, structural dynamics in severe environments, and high-velocity impacts. The company designed advanced high-explosive systems, including ordnance devices and large-scale experimental simulators. Its affiliate, California Computing Resources, of Chatsworth, California, operated a computer service bureau.

Finally, Meteor Communications, of Kent, Washington, was a communications technology and manufacturing company with expertise in meteor ion-trail communications.

While Titan's leaders were busy adding these businesses to the company's lineup, Rolf Erikson, who by the time he retired in 1990 was Titan's vice president of corporate development, began selling those concerns that didn't complement the company's strategic goals. "It was a classical asset redeployment," said Erikson. "Many of EMM's assets were tired, so we sold off those pieces and took that money to buy new pieces—like playing a game of chess."[36]

Indiana General Ferrite Products, based in Keasbey, New Jersey, was sold to Allen-Bradley, of Milwaukee, a division of Rockwell International, in October 1985. The company, which manufactured parts for large industrial motors, accounted for less than 10 percent of Titan's revenue.[37] Indiana General Motors Products, based in El Paso, Texas, was sold to Century Electric in June 1986.

"The coming together of Titan and EMM combined different cultures and created some very interesting and exciting challenges, which included, of course, divestitures and plant closures," said Marsh Nelson, who retired as senior vice president, secretary, and general counsel in 1996. But, he observed, Titan's transition away from the large number of offshore manufacturing facilities it had acquired in the EMM merger helped "lay a foundation for the later, more important transition from the defense business into the commercial world."[38]

Streamlining

With key personnel in place and acquisitions and divestitures in the works, Titan's leaders decided it was time to streamline management and operations and reorganized the company into three new business groups: Titan Technologies, Titan Electronics, and Titan Systems.

Titan Technologies, led by Gerold Yonas, was headquartered in La Jolla, California, and explored and developed emerging technologies. The group was involved in development and application of large computer programs for the modeling of scientific and engineering problems; development of electro-optical technology for sensor applications; and development, design, and manufacture of systems in the field of pulsed power, such as microwave sources and advanced simulators. Pulsed power technology made it possible to gather energy and release it in short, powerful bursts.

Kenneth Driessen, former vice president of IBM's Federal Systems Division, joined the company to become president of Titan Electronics. This group continued to produce militarized, embedded computers and computer memory products for both military and commercial use. It also provided engineering and software for advanced designs in satellite navigation systems, as well as design, development, and manufacture of meteor-burst, long-distance radio communication equipment. Titan, in fact, was considered the world's leading supplier of core memories for use in adverse environments.

Titan cofounder Jack McDougall became president of Titan Systems, also headquartered in La Jolla. Titan Systems specialized in system development and integration for C^3I systems and performed system engineering, integration, and software development for defense, space, and electro-optical systems. Titan Systems worked closely with Titan Technologies and Titan Electronics to develop products that could be integrated into larger systems.

In October 1986, Titan Systems received the DoD's prestigious Cogswell Award for excellent security performance. Titan was one of 18 companies to receive the award out of 12,500 DoD contractors nationwide.[39]

Meanwhile, Titan cofounder Ed Knauf was promoted to the newly created position of executive vice president of The Titan Corporation. Based in Washington, D.C., Knauf was responsible for developing Titan's strategic business plans and for identifying and pursuing large governmental programs.

Divisional Developments

Busy with mergers, acquisitions, divestitures, reorganization, and streamlining, The Titan Corporation was in a steady state of transition during 1985 and 1986. Still, nearly all of the movement was positive, and the divisions that had previously been in place continued uninterrupted.

Titan's computer division, SESCO, made a move to enhance worldwide market penetration of its SECS computer product line in June 1985 by entering into a licensing agreement with British Aerospace Dynamics Group's Bristol Division. British Aerospace planned to use the computers in its weapon and aircraft systems and planned to market SESCO products in the United Kingdom.

SESCO, headquartered in Encino, California, held an exclusive license with Intel to produce military-grade equivalents of Intel's iSBC line of

As Titan continued to bring in new talent, cofounders Ed Knauf (enjoying the celebration of the EMM merger) and Jack McDougall took on new roles. Knauf became executive vice president of The Titan Corporation, and McDougall became president of Titan Systems, the company's defense systems segment.

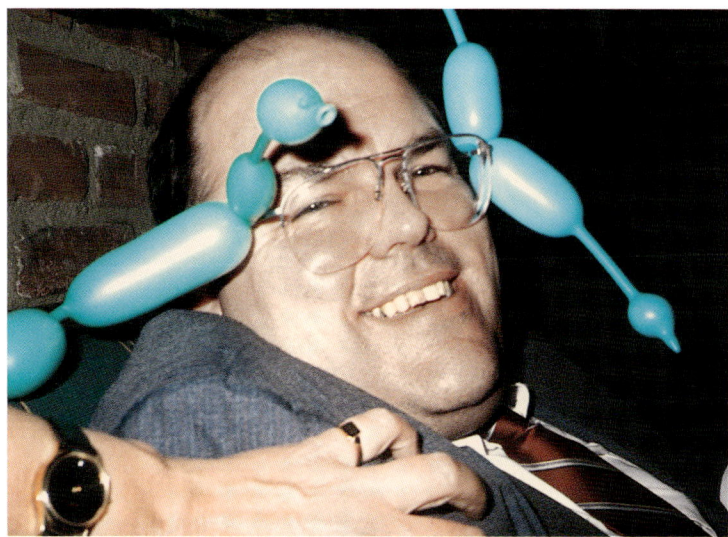

Top: Titan's technological expertise was essential to SDI and applied to such important programs as the Strategic Pilot Associates Program and the Smart Weapons Program.

Above: Titan specialized in developing and integrating hardware and software for military applications. Chief engineer Henry Young (right) displays a radiation-hardened computer that acts as a U.S. Navy link to the Milstar military satellite communications system. Systems analyst Dr. Frank Bletzacker (left) holds a core memory system designed for military aircraft.

single-board computers.[40] The division's combined 1985 sales of computers and memories totaled about $33 million.[41]

In October 1986, SESCO won a $3.75 million contract to supply core memory systems for the Air Force's F-15 aircraft.

Titan's Computer Power Products, based in Gardena, California, had a $1 million contract with the Department of Energy's Mercury, Nevada, site for rotary uninterruptible power supply (RUPS). During underground tests, Computer Power's RUPS systems were the only ones able to survive the ground shock wave. They were also chosen because of their ruggedness and ease of installation. Despite this victory, Titan chose to discontinue and sell Computer Power Products in 1986 because the company was not a good fit with Titan's strategic plan.

Canada Alloy Castings, a high-alloy castings company based in Canada, was another less-than-desirable fit for Titan. Erikson began looking for a buyer, which he would find in July 1987.

The Payoff

By the end of 1986, Titan was the 10th-largest public company in San Diego. (As a testament to Titan's strength, by 2001 all but Titan and one

Titan engineers Royce Bowyer (left) and Duaine Berger (right) pose with the rotary uninterruptible power supply (RUPS), which protects electronic equipment from power spikes, surges, and outages.

other of those 10 companies, Cubic Corporation, would disappear.[42]) Revenue for the year increased to $121.4 million, up 23 percent over 1985. Since mid-1985, the company had sold or discontinued operations that represented approximately $35 million in annual revenue. Income from continuing operations was $6.6 million, or $.46 per share, compared to $1.5 million, or $.04 per share, in 1985. With a $4.9 million loss from the discontinuance of Computer Power Products, net income for the year was $1.1 million.[43]

On the whole, Titan's employees were well satisfied with their work. In 1985, Dr. Martin Fricke, who headed Titan's advanced research, told *Barron's* magazine, "I've worked at national labs, at universities, and at larger companies, and I've never been as happy as I have been here."[44]

Analysts were also pleased with the new company. Titan has "a base for growth in excess of 20 percent annually," reported McKewon and Timmins, also noting that Titan's management was "very strong."[45]

"Our primary corporate objective is to build a major corporation that will be a world leader in developing and providing high-technology systems and products to meet the needs of a wide range of governmental, industrial, and international customers," Ray wrote in Titan's 1986 annual report. "We are well on our way with a company that has a management team, corporate structure, financial strength, and technical capability to support continued significant growth."

By all appearances, The Titan Corporation was on the fast track to becoming a huge success story. No one—not executives, employees, or business analysts—could see the clouds that loomed on the horizon, but as Kavanaugh observed years later, "You could always count on Gene Ray to see his way through the black clouds."[46]

When defense spending shrank, Titan moved further into nonmilitary marketplaces. Using existing technology, it began developing Doppler radar systems for the Federal Aviation Administration to warn aircraft of wind shear, or sudden changes in wind speed or direction.

CHAPTER FOUR
A SHIFTING WORLD
1987–1989

Contracts are being canceled, delayed, and stretched out. I don't think there has been any major defense contractor that has not felt the effects of this uncertainty.

—Gene Ray, 1988

IN 1987, THE TITAN CORPORATION faced a possible derailment on its fast track to success when the United States and the Soviet Union, fierce enemies for some four decades, took the first steps toward a more peaceful relationship. It was a move that would ultimately bring longed-for peace between the two nations, but it would have a serious impact on the defense industry. As one New York business analyst conjectured, "I think there's a real fear [among investors] that peace may break out."[1]

Only four years before, in 1983, President Ronald Reagan had labeled the Soviet Union "the evil empire." Such a foe warranted heavy Star Wars spending, but in 1985 the Soviet Union gained a leader who would not only transform that country but change the course of world history. Soviet Premier Mikhail Gorbachev, wishing to focus on the frail Soviet economy, adopted "glasnost" (open dealings at home and with the West) and hoped to put an end to the costly arms race.

Gorbachev and Reagan met several times before reaching an agreement. At the 1985 Geneva summit, when Gorbachev called for scrapping SDI, Reagan recommitted to its development. A 1986 summit between Gorbachev and Reagan in Reykjavik, Iceland, also ended in stalemate. But by their third summit, in Washington, D.C., the two were ready to agree.

Reagan was partially motivated by a Democrat-controlled, cost-cutting Congress. The 1985 Balanced Budget and Emergency Deficit Control Act, better known as the Gramm-Rudman Act, called for an end to the U.S. budget deficit by 1991, and Congress was seeking cutbacks across the board. An especially ripe target was SDI, with its multi-billion-dollar development costs.

Reagan and Gorbachev signed the INF (Intermediate Nuclear Forces) Treaty in 1987, effectively eliminating land-based, intermediate-range nuclear missiles such as SS20s and cruise missiles.

It was a first step toward the end of the Cold War.

Rocky Times

In terms of its workload and contract backlog, Titan was not immediately affected by this peace overture. The company had created a solid base of work, and more contracts were rolling in. However, word of the impending treaty took its

---

Among its many communications products and services, Titan offered advanced navigation simulation products for geosynchronous positioning and communications.

toll on the stock market. Titan's stock, which had risen to a high of $11.88 in February 1986, was down to $6.63 in April 1987.²

Ray realized just how far the defense industry was sliding after a conversation with Andy Gaspar, the financial analyst that Ray had worked with at Warburg Pincus when he first contemplated taking Titan public. By then Gaspar had become a personal friend of Ray's.

"Gene and I compared notes about the future of the defense industry," Gaspar remembered. "We talked about the downward cycle, how I did not think it would be reversed anytime soon."³

"He laid out how bad it was going to be," Ray said. "I thought he was totally exaggerating, but he scared me badly enough that we started taking action. It turned out to be worse than what he said."⁴

Some business reporters tried to sum up the situation. "Not so long ago, defense electronics companies were the toast of Wall Street," wrote a reporter for the *Los Angeles Times*. "These days, defense electronics stocks, including Titan Corp. of San Diego, more closely resemble stale beer."⁵ Or, as defense securities analyst Thomas Lloyd-Butler said, "Titan's fortunes are wedded to congressional appropriations, and prospects for defense spending over the next several years are not all that rosy."⁶

"We were intimately involved in so much more beyond Star Wars," said Ed Knauf. "But oftentimes [for security reasons] we couldn't tell the public the exact nature of some of our business."⁷

Potential defense cuts were not the only challenge Titan would face that year. Like all U.S. businesses, the company was jolted by the plunge of the stock market in October 1987. Black Monday, as the date came to be known, sent the company's stock price down almost 50 percent. But Titan chose to take a positive approach. Titan's leaders immediately initiated a common stock repurchase to take advantage of low prices. Then they took a hard look at the company's efficiency and introduced a company-wide cost reduction program.⁸ This proved to be a prudent move. In January 1988, Congress made its position clear. Military spending on strategic technology would be frozen at its current level of about $292 billion a year.⁹

"The shoe has to drop at some point," said Titan Electronics President Kenneth Driessen. "There are a lot of areas that we could spend more to develop, but looking at the economics and the potential impact of a spending cut, we've decided to hold off."¹⁰

But perhaps Ray summed up the circumstances best when he said, "The overall defense environment is damn tough."¹¹

Advancing Cautiously

While the difficult environment slowed Titan's leaders, it didn't stop

Titan specialized in systems integration. In 1988 it won a contract with the U.S. Navy to update the processing of communications data.

Titan's technology helped train U.S. Marines through simulations software.

them. Some defense contractors decided to try to ride out the budget cuts, lowering their revenue at the same time. Others put themselves on the market, becoming part of a mass industry consolidation. Titan's leaders strategized cautiously and shrewdly, choosing to move the company's technologies, many of which had been funded by the government, into a commercial marketplace and developing patents and information technology (IT) along the way. The company planned to continue with its defense work, but it would go after more conventional contracts (not SDI-related) and seek niches that would be safe from cuts. The company would pour more resources into finding commercial applications for its defense technologies. It would also be open to technologies that were not defense related at all.

The shift in business would take time, but Titan's existing and new defense contracts promised to carry it through the slump. Several contracts were awarded to Titan in 1987 and 1988:

- A $3.2 million contract from the U.S. Naval Training Systems Center to provide Universal Maintenance Training Systems (UMTS) to the Marines at Camp Lejeune in North Carolina.[12]
- Five new contracts totaling more than $40 million, awarded in April 1987. The largest came from the Army's Strategic Defense Command and involved C^3I and battle management as well as systems engineering and analysis. A second involved a team effort to provide engineering, installation, integration, and test and support services for the Office of the Secretary of Defense's Crisis Coordination Group facility. Two contracts were with the Defense Communications Agency and included engineering, analysis, and technical input to C^3I systems. The fifth contract was for baseline analysis of command and control in selected U.S. Air Force command centers.[13]
- A $6.4 million contract from Martin Marietta Energy Systems for a baseline requirement analysis of command and control in selected U.S. Air Force Crisis Management centers. This work supported the Air Force's World-

wide Military Command and Control Information System Program.[14]
- A $3.5 million contract for specialized computers and software to support the initial operational evaluation phase of the Fiber Optic Guided Missile (FOG-M) program. Titan was to integrate both the FOG-M gunner's computer and the Multi-Missile Fire control unit into the overall system.[15]
- A $3.9 million contract from the Naval Weapons Support Center.
- Two Defense Nuclear Agency contracts totaling more than $10 million. The first covered a project aimed at increasing the effectiveness of earth-penetrating weapons. The second entailed the analysis of environments produced by multiple nuclear explosions.[16]

Titan won at least one major nondefense contract during this period, with Raytheon in November 1988. Titan worked with Raytheon to develop 47 Terminal Doppler Weather Radar (TDWR) systems for the Federal Aviation Administration (FAA). The radar warned aircraft of wind shear, a sudden change in wind speed or direction that can cause a plane to rapidly lose altitude.[17]

Despite these contracts, Titan experienced some wind shear of its own. In 1988 alone, six of its contracts were canceled due to changing defense priorities.[18] Profits began to slide. At a May 1988 shareholders meeting, Ray was straightforward about the company's concerns. "Just like you, your management team is disappointed in our recent financial performance," he told shareholders. "What it boils down to is the fact that we are going to have to be more competitive."[19]

Chairman Webb conceded that Titan "has too many problems. Some of the problems are our fault and some are beyond our control." To resolve the problems, Webb said, "We've done everything short of consulting an astrologer."[20]

Belt Tightening

Titan didn't resort to casting horoscopes, but it did initiate some serious restructuring and cost-cutting efforts. Employment levels were reduced, corporate costs were cut by 14 percent, and the company's travel budget was slashed by $300,000. The company also planned to build a new headquarters in Sorrento Mesa, a less costly area of San Diego, which was expected to cut the company's real estate cost by 30 percent.[21]

In addition, Titan sought to divest businesses that were hurting its bottom line. In July 1987, it sold approximately 80 percent of its interest in two small Canadian subsidiaries, Canada Alloy Castings and Canada Investment

Titan's Optical Air Data System (OADS) accurately measured an airplane's speed and local freestream pressure and temperature and replaced less precise pneumatic air data systems. In 1989, Titan demonstrated its prototype OADS equipment aboard an F-16 fighter.

Castings. "The sale of these two entities is part of our overall strategy of reducing our interest in operations that don't fit our primary business and using those resources to better long-range advantage," Ray explained.[22] Titan's remaining 20 percent interest in the subsidiaries was sold in October 1988.

In January 1989, Titan announced the sale of its Advanced Materials division for $15.5 million, as well as its 87.5 percent equity interest in CAST. Advanced Materials, which produced ferrite powder for photocopy machines, went to Nippon Iron Powder.[23]

A more important divestiture was the sale of Meteor Communications (MCC) in June 1989. Titan had acquired MCC in 1986. The Seattle-area company had developed a technology for communicating by bouncing radio signals off the tails of the millions of meteorites that enter the earth's atmosphere each day. While innovative, the technology wasn't being used and was therefore losing money. "It was a communications Edsel," Ray said. MCC was sold to five Titan employees.[24]

Business analysts approved. "I give these people credit for recognizing early on that it doesn't make sense to allocate resources to this type of business," one analyst told the *San Diego Business Journal*. "It really doesn't fall into the types of things Titan bids on."[25]

Careful Growth

While divesting businesses that were poor performers or that lacked synergy with Titan's main focus, the company also made some strategic acquisitions that would allow it to offer more complete project capabilities.

In March 1987, it purchased Pulse Sciences Inc. (PSI), of San Leandro, California. PSI developed, designed, and manufactured high-power microwave systems and strengthened Titan's technology base of high-power microwave sources, advanced simulators, laser power conditioners, and advanced accelerators. It also made possible new commercial applications such as radian processing of materials.[26]

In July 1987, Titan acquired Advanced Digital Systems (ADS), a San Diego–based company that

Led by Sid Webb, chairman, and Gene Ray, president and CEO, Titan survived some of the defense industry's most difficult years by applying its existing defense technology to commercial applications.

provided software and systems engineering for U.S. Navy satellite communications.

Another company acquired in 1987 was Physics Applications Inc. (PAI). PAI added significantly to Titan's advanced conventional ordnance business.

At a May 1987 shareholders meeting, Sid Webb discussed Titan's acquisition strategy. "We're not a big conglomerate trying to gobble up companies," he said. Rather, Titan was seeking small to mid-size firms that were a good fit and could boost overall sales. Titan had considered acquiring two larger, publicly traded companies with sales between $80 million and $90 million but had held back. "The risks were too great in the long term," Webb told shareholders.[27]

He and Ray summed up Titan's acquisition strategy in the 1987 annual report.

The principal thrust ... has been to obtain companies that enable us to offer customers

CHAPTER FOUR: A SHIFTING WORLD

Titan's patented explosive detection with energetic photons (EXDEP) technique used radiation analysis to detect explosives. It could be used for enhancing airport security and for other antiterrorist applications.

complete project capabilities, from systems design and technology development to fabrication, manufacturing, and installation. This, in turn, gives us the capability to continue to pursue and win larger contracts.

Some years later, Ray expounded further on this philosophy. "The first thing is to decide where you want to go and what your strategy is for getting there, then look for things that fit that strategic picture," he said. "Second is a reasonable cultural fit. Every company has its own culture and its own personality, but, boy, they had better be reasonably compatible. If they're not, it's very difficult to get them to work together."[28]

Despite the challenges Titan had to overcome, in 1987 the San Diego chapter of the Association of Corporate Growth acknowledged the company for its "outstanding sales and earnings over the last six years."[29]

A Sharper Focus

Titan was by no means planning to get out of the defense business, but its leaders knew the company needed a sharper defense focus to survive budget cuts. They chose to bolster C³I, an area to which Titan had already given prominence. As the *San Diego Union-Tribune* noted in October 1988, C³I was "one of the few bright spots in an era of Pentagon belt-tightening."[30]

"As a general categorization, I would say, yes, it is going to be a relatively attractive area for funding support in an era of otherwise lean defense spending," a defense electronics analyst told the *Union-Tribune*. "I would be surprised to see this area suffer as much as other areas of the defense budget."[31]

The main attraction of C³I systems was their promise as "force multipliers," which could help stretch the value of existing resources. It was an appealing concept to a Congress intent on slimmed spending.[32]

In addition, Titan began to go after more contracts to design and build "critical components" for tactical weapons. These tactical weapons included torpedo warheads, charge-resistant armor, and electromagnetic mine detection.[33]

Exploring New Technologies

Titan also began exploring new technologies. To facilitate its strategy to diversify, Titan's leaders put together a three-person committee in 1989 to go to every Titan office and evaluate what concepts might be built into commercial businesses. The committee identified 54 different technologies, products, and concepts that had potential for commercialization.

In early 1989, the company began testing a new, proprietary, patented technique for detecting nonvapor explosives, a process that could be used to discover unexploded artillery, terrorist devices, explosive mines, or weapons in airline baggage and cargo. The new technique was called EXDEP, or explosive detection with energetic photons.[34]

Because Titan's focus had shifted away from his specialty of SDI technologies, Gerold Yonas decided to return to Sandia Laboratories in May 1989.

Brief Interlude

Though Titan reported an $11.2 million loss for 1988,[35] the corporate restructuring and new strategies steered the company in the right direction. In 1989 the company won a number of new contracts:

- A $3.8 million contract from the Space and Naval Warfare Systems Command to produce a Shipboard Communications and Computer Access Terminal.[36]
- A $7.6 million contract from the Naval Ocean Systems Center to provide engineering support for the integration of surveillance and C^3I systems.[37]
- A $5.3 million contract to provide life-cycle support for Navy Fleet Satellite Communications Systems.[38]
- An $11 million contract from the Naval Surface Warfare Center to provide research

Titan began working on a new supercomputer project for the U.S. government called Direct Solution of Turbulence (DST), which reduced the effect of air and water turbulence on aircraft and submarines, such as this Trident submarine.

and development for a high-power electron beam accelerator concept.[39]
- A $1.3 million contract from the Office of Naval Research for the initial phase of the Direct Solution of Turbulence (DST) program, which was designed to reduce the effect of air and water turbulence on aircraft and submarines.[40]

While these new contracts appeared promising for Titan, any confidence they produced would be short lived.

Another Jolt

While Titan had been reorganizing and refocusing, the troubled Soviet economy was gasping for breath as a series of revolutions swept away the Communist regimes that had governed Soviet satellite nations in Eastern Europe. Then in November 1989, the Communist government of East Germany collapsed, and Germany was reunified. The Berlin Wall, the most prominent physical symbol of the Cold War for more than 28 years, came down. Much of the world, including the United States, celebrated, but for the defense industry it was another major jolt.

"It's not good for our business, but it's what our country has been trying to accomplish for years," Ray commented. "And we've done it without firing a bullet."[41] He noted that Titan would pull through for the next few years because the company had its highest backlog of contracts yet. "How we do after that depends on how well we diversify," he said, adding that Titan had 54 products it could apply to the commercial market.[42]

It was a good sign when Titan won a major nondefense contract late in 1989. The contract, with a potential value of $6.3 million, was to support the Next Generation Weather Radar (NEXRAD) program for the Department of Commerce.

Even so, huge challenges lay ahead. Major defense cuts were a certainty, and the U.S. economy was headed for a major setback. Titan would soon be forced to prove its mettle.

As Titan diversified, it applied its technology to education, offering management and technical consulting, product support, and systems integration services for grades K–12 and higher education institutions.

CHAPTER FIVE
MEETING THE CHALLENGE
1990–1992

It's the most uncertain time that I've ever seen in the defense industry. You don't go and build a new building on expected sales, you don't plan on making big purchases that aren't necessary, and you keep a much closer eye on what Congress is doing.
—Gene Ray, 1990

AS THE 1990S UNFOLDED, Titan's leaders faced the greatest test in the company's 10-year history. They needed to move swiftly to reposition the company, but Titan's revenue was dependent on a precarious defense budget.

One thing was clear: The next few years would be challenging.

Some in the defense industry had felt hope when President George Bush named Dick Cheney secretary of defense in 1989. As a Wyoming congressman, Cheney had been a steadfast supporter of SDI and a variety of other military growth initiatives. But after being named to his new post, Cheney faced a quandary. The Cold War was winding down, and Congress and the president were calling for major military cuts.

As Cheney would say in 1991, "We're in a position where, as secretary of defense, I've got a 10-pound defense program and a six-pound defense budget. Somehow, we've got to squeeze one down so it fits."[1]

When the Soviet Union collapsed, Cheney, too, began to view the defense budget cuts as a serious necessity, but he was concerned that nations such as Iraq, Iran, and North Korea would be able to acquire nuclear weapons from the newly independent republics. Cheney called the cuts "very, very painful."[2]

Some defense contractors were also pained. As the *San Diego Union* reported in January 1990, "After years of prosperity driven by unprecedented Pentagon procurement, local defense contractors privately are questioning whether they can survive the fallout of détente between the United States and the Soviet Union."[3]

San Diego County was the fourth-largest recipient of defense contracts in California, and the decline had forced large San Diego contractors such as Rohr Industries, Cubic Corporation, General Dynamics, and TRW to lay off thousands of workers.[4] Titan had also cut staff, an experience that Ed Knauf, Titan cofounder and executive vice president, painfully remembered as being "very tough."[5]

Still, some business analysts remained bullish about Titan. With the company's stock selling for about $2 a share, some investors thought the company was "headed for bankruptcy," said analyst Larry Selwitz of Cruttenden and Company. In Selwitz's opinion, however, Titan was "thriving."[6]

Titan's very large scale integration (VLSI) chips were capable of a wide variety of communications processing functions, including digital signal processing, forward error correction, global memory interface, and secure communications.

"Several technologies developed in the last three years put Titan in a strong position to win major contracts during 1990," he said. "Many of its current contracts, generally in the $1 million to $10 million range, have the potential to lead to follow-on or next-generation business of 10 times current funding levels."[7]

Indeed, when Cheney announced a $34.6 billion cut in the DoD's $200 billion aircraft program in April 1990, Titan officials told the San Diego Tribune that the proposed cutbacks would have little effect on the company's business. The changes, they said, would hurt "metal-bending" defense contractors but not those who specialized in "smart" technologies.[8]

Titan's leaders were increasingly focused on broadening technologies and moving them into new, commercial uses, both in the United States and abroad. In the 1990 annual report, Gene Ray and Sid Webb announced that expanded marketing efforts had brought the company work from Sweden, the United Kingdom, France, Germany, and Saudi Arabia. In the United States, nondefense customers included the Department of Energy, the Drug Enforcement Administration, and NASA. Titan was also moving into medical technology, weather forecasting, specialized computing applications, and education.[9]

Ray and Webb were optimistic in their letter to shareholders, noting that the company's professional expertise, its principles, and "plain hard work . . . should continue to serve us well as we go forward to meet the challenges of the next decade."[10]

Bringing in Linkabit

In 1990, when President Bush called for economic sanctions against Iraq and began Operation Desert Shield, there was hope on Wall Street that the U.S. military buildup in Saudi Arabia would boost the defense business. Reflecting those hopes, defense industry stock prices rose.

Aerospace Daily reported that Titan was told to "send all the satellite ground terminals you can, and [the U.S. military will] send trucks to pick them up."[11] The timing was fortuitous for Titan since it had just finalized a deal to purchase

The Next Generation Weather Radar (NEXRAD) system provided early detection of weather phenomena such as hurricanes, pictured here. In support of NEXRAD, Titan provided oversight, deployment, operations, training, hardware and software systems engineering, logistics, planning, and quality assurance.

CHAPTER FIVE: MEETING THE CHALLENGE

Above: At the end of 1992, Gene Ray and Sid Webb were able to announce with pride that, despite the continuing downturn in defense spending, Titan had grown in revenue and earnings per share for four consecutive years.

Right: Titan's single channel transponder system received critical command and control communications transmitted by defense satellites. The system was portable and could be moved easily and set up quickly.

M/A-COM's subsidiary, Linkabit Corporation, a San Diego–based company that manufactured earth-based satellite communications equipment for the U.S. government.

Linkabit had been formed in 1968 by two engineering professors, Andrew Viterbi and Irwin Mark Jacobs. (Viterbi and Jacobs also founded Qualcomm, and Viterbi was renowned in the field of electrical engineering for developing the convolutional coding named after him.) The small company pioneered a digital signal processing technology for satellite communication and in 1980 had become a subsidiary of M/A-COM.

In 1988, Linkabit began preparing to compete for a program for the U.S. Navy to develop Mini-DAMA (demand assigned multiple access), ultra-high frequency (UHF) satellite terminals. The DAMA technology, in the words of Linkabit's senior vice president, Tom Trimble, "allowed users to digitize and multiplex a lot of conversations in the same pipeline."[12] The contract required that Linkabit partner with another company to give the Navy a second source from which to get the modems. Linkabit began looking around for that second source and hit upon Titan.[13]

It was after the two companies had won the Navy contract in July 1989 that Ray initiated discussions to buy Linkabit. Titan acquired the company a year later, renamed it Titan Linkabit, and moved Titan's headquarters into Linkabit's facilities. Titan also sold the Linkabit buildings and leased them back, the proceeds of which reduced Titan's debt.

"Linkabit's technology was a brand-new type of work for us," said Jack McDougall, by then executive vice president. "We had been mainly

doing systems analysis, systems engineering, and the EMM manufacturing and computer technology, but Linkabit was a very high-tech, hardware-oriented company. It required a lot of [research and development] to nurture the product and keep it growing into new directions."[14]

And Titan certainly kept Linkabit's development going. Over the years, Linkabit became, in the words of Linkabit's Bill Zettinger, vice president of programs, "a technology engine for Titan," specializing in developing communications technology, modems, and signal products.[15] Furthermore, Linkabit helped move Titan into more government projects.[16]

While Titan was able to assist the United States' offensives in Desert Shield and Desert Storm, the role did not last long. The United States began its air attack on Iraqi forces in Kuwait on January 17, 1991, and then on February 24, 1991, launched what was to be a very short ground war. By March 3, Iraq had agreed to a formal truce.

Though the war was short, Titan felt it had won points with the U.S. government because the battle had emphasized the need for advanced communication systems—exactly the type Titan specialized in. In all, 11 different Titan information products were deployed during the war.

Strategic Alliances

While participating in the war effort, Titan was also fully engaged in diversifying its business interests. The company wanted to hold on to what it had but was evaluating how to expand into other areas.

Two alliances formed during 1991 exemplified Titan's dual intentions. First the company formed a joint venture with Ericsson Radar Electronics to create GBS Joint Venture, aimed at winning the U.S. Army contract for the Ground Based Sensor. The two companies were able to pool resources to go after a contract that could be worth as much as $1 billion. Together they were able to compete against such big guns as Lockheed Martin, Marietta, and Hughes.[17] Going after the contract was no small undertaking. The proposed hardware that Titan and Ericsson assembled weighed 1,950 pounds and had to be flown to Washington aboard a chartered jet.[18]

Titan produced the LST-8000 satellite terminal, which provided secure worldwide communication. Despite its size and range, the terminal could be carried in five cases and set up by two people in less than an hour.

Unfortunately, Titan ultimately lost the contract to Hughes because of cost.

Second, Titan Linkabit formed a strategic alliance with the Motorola Government Electronics Group to produce a low-cost, high-performance satellite terminal for a wide variety of secure communication applications. By working together, Titan and Motorola could cut in half the time it would take to develop a new satellite communication terminal. Titan wanted to employ the technology for nonmilitary uses and brought in new employees with the specific experience needed to take the satellite terminal into the commercial marketplace.

Acquisitions

Titan also continued its strategic acquisitions program. In August 1991, the company purchased the Stonehouse Group, a Denver-based company that provided software consulting and software tool development to the U.S. government. Stonehouse also had offices in Vienna, Virginia. The acquisition expanded Titan's capabilities in information and database systems.[19]

Then in January 1992, Titan acquired the Satcom product line from Gamma Microwave, a Santa Clara, California, company. The Satcom line consisted of commercial satellite transmitter-receivers needed for the installation of commercial satellite terminals both in the United States and abroad. The acquisition was another step in expanding Titan's business into the commercial arena.[20]

New Contracts

As Titan continued its push into the nonmilitary business world, new contracts began to materialize. In March 1991, the company won three contracts with a total value in excess of $8 million:

- A contract to build a 200-million-electron-volt linear accelerator for Brookhaven National Laboratory on Long Island. This turnkey system would be used to accelerate electrons to precise energy levels, a technology with applications in x-ray lithography for the next generation of microchips.
- A contract to develop a mobile microwave transmitter system for Saab Aircraft of Sweden to test electronic equipment at Saab's manufacturing facility.
- A contract to build a specialized, high-power, pulsed modulator for use in the aerospace industry.

In May 1991, Titan Electronics won a $6 million contract from ITT Federal Services for short-range radar control groups. It was the largest contract in the division's history. Then in December, Titan landed a $12 million contract from the National Laboratory of Frascati, Italy, to build the particle source for a new, high-energy nuclear accelerator. The particles were to be used to test fundamental hypotheses about the composition of matter and the symmetries of the universe.

Titan remained in the defense contract game as well. In March 1991, Titan Engineering Services won a $19 million award to provide services to the Pacific Missile Test Center at Point Mugu, California. To win the contract, the company teamed with Comarco and Mantech in yet another effort to offset rising costs, get new distribution channels, and reduce competition.

Titan continued its work in pulse technology. This 6-million-volt, 400,000-amp ion-beam pulse generator was used for research on nuclear fusion for electric power production.

Titan's patented explosive-detection technology could detect mines or other explosives by using an electron beam x-ray machine similar to those used in medical facilities for cancer treatment.

A Boulder Rolling Downhill

In August 1991, when Ray was invited to write a guest editorial in the *San Diego Union* about the unraveling of the Soviet Union, he was pragmatic. But he still made a pitch for more defense spending. "You can't stop a boulder going full speed down a steep hill," he wrote. "Your best bet is to adjust its angle of descent."[21]

Ray said he hoped for "a smooth and economically successful transition to a democratically elected government and free and open society" for the former Soviet Union. Yet he cautioned that such a government might not be the outcome of the Soviet Union's passing and that the United States needed to stand ready for possible "chaos."[22]

New Developments

Even while Ray encouraged greater defense spending, Titan did not depend on it; in fact, the defense budget continued ratcheting downward. In October 1991, the company created a new division, Titan Information Systems, to increase its development, manufacture, and marketing of commercial products. The new division was to concentrate on the areas of television encryption, satellite telecommunication systems, and information technology services.[23]

Ray announced the new unit at a meeting of the American Electronics Association. "We have existing patents which we are now adapting to the broader commercial market," he said. "We have development and production expertise which we will utilize in this new arena."[24]

In September 1992, Titan took another step toward diversification by licensing Messerschmitt-Bolkow-Blohm in Munich, Germany, its patented rights to an explosives detection technique. The system was designed to improve airport security in Europe by helping locate and identify explosives in luggage. This represented the first time Titan was able to implement its strategy of bringing value to shareholders by licensing its technology.[25]

Electron-Beam Sterilization

After three years of work, including outside market research, Titan made a pioneering move in early 1992 by creating a new commercial business from an existing technology, electron-beam sterilization. This technology had evolved from linear accelerator technology originally designed to study the effects of nuclear weapons.

The linear accelerator technology actually stemmed from one of Titan's acquisitions back in 1983. Gary Loda, who founded Beta Development, had developed highly sophisticated machines to test the effects of nuclear weapons in laboratory environments. A few years later, Tony Zante picked up where Loda had left off and recognized that the high-energy electron beams could be used to kill bacteria without harming products.

Meanwhile, a leak occurred in Atlanta at a cesium 137 plant, which used a radioactive, potentially hazardous method to sterilize medical devices. The Department of Energy (DOE), which leased cesium plants to various companies throughout the country, subsequently recalled all its leases.[26]

Titan's initial goal for the electron-beam sterilization system was to provide an alternative to existing systems that used radioactive materials or hazardous chemicals. High-energy electron beams killed bacteria on medical devices, but unlike the chemical processes it replaced, electron sterilization did not damage the products or leave residue. It promised to replace sterilization by ethylene oxide gas, which killed microbes but was a known carcinogen.[27]

The board gave the go-ahead in June 1992, and Titan began building an electron beam sterilization facility in Denver. The unique facility would be operated by the newly formed Titan Scan division as a contract service operation. Brian Williams, a Ph.D. physicist, became Titan Scan's first employee. Tom Allen, who later became Titan's vice president for systems engineering integration, was hired for his expertise in the commercial sector. Gary Pageau, who had experience with cesium 137, became head of marketing for the division, and Ray Calhoun, a certified public accountant who had worked with Ray at SAIC, became Titan Scan's CFO. Together these four men did much of the early development of electron-beam sterilization technology.[28]

"We spent tens of thousands of hours in engineering to develop the basic concept, the control systems," remembered Allen. "And out of that effort came the base patents that we rely on today."[29]

At the core of the Titan sterilization method was a process later called SureBeam®, which the company patented in 1993. This technology used a 10-million-volt, 10-kilowatt accelerator to generate electron beams that could sterilize medical products in their original packaging.[30]

Just one month after it had announced it would build the Denver facility, Titan received two contracts totaling more than $3.5 million to sterilize petri dishes, specimen containers, and culture tubes for two Denver-area medical product manufacturers. By the time the facility opened in June 1993, the company had contracts worth $5 million. Plans were made to open two more such facilities in San Diego and Chicago.

"We're very optimistic about where this line of business can take us," Ray said, noting that the U.S. market for medical sterilization totaled $300 million annually.[31]

Little did Ray know how well founded his optimism was. Over the years, SureBeam would take Titan into markets not even dreamed about in 1993. Just one year later, in 1994, Titan began looking seriously at SureBeam's potential for destroying bacteria and other microorganisms in food. More dramatic uses would follow in years to come.

More Defense Work

While Titan was making these successful moves into commercial diversification, it continued to win defense contracts. In the fall of 1992, the company received a $2.6 million award from Hercules Defense Electronics for 550 sets of Titan's proprietary SECS 80 series of computer and memory modules for installation in the Hercules AN/AAR-47 missile warning system. The system provided slow-moving, low-flying aircraft such as helicopters and fixed-wing planes with an electro-optic detection system.

Titan's accelerators provide high-energy particles that are used for medical research, explosive detection, and scientific investigation. In a few years, the technology would spark other innovations as well.

Titan's Severe Environment Computer System (SECS) 80 was rugged enough to qualify for a number of military applications, including providing slow-moving aircraft such as helicopters with an electro-optic detection system.

In December 1992, Titan won a $6.1 million contract from General Electric Aerospace to develop a signal processing subsystem and a Milstar satellite simulator. The subsystem was part of a communications terminal being built by GE Aerospace for the Army.

Success in Diversification

By the end of 1992, it was clear that Titan's diversification strategies were working. Both defense and commercial contracts were rolling in. In 1990, revenue had risen 27 percent, from $97.7 million in 1989 to $124 million. Revenue swelled again in 1991 to $146.5 million. In 1992, the company experienced flat sales and earnings of $149 million, but commercial sales rose to $28.3 million, a $10 million increase over the previous year. The company's client base at the end of 1992 was 69 percent defense, 12 percent government, and 19 percent commercial.[32]

An article in the *San Diego Business Journal* in December 1992 highlighted Titan as one of the San Diego defense companies that was finding success in diversification. In that article, Ray commented on the necessary elements for conversion from defense to commercial.

"You need three things," he said. "A sellable product and someone who knows how to sell it in the marketplace, management with commercial experience, and capital. Defense, at best, has two of those. And when you start out with a two-legged stool, you're doomed for failure." What many defense companies lacked, said Ray, was the commercial know-how. Titan, knowing it lacked that basic ingredient, had hired people who had it.[33]

"Titan figures it has all three legs of the stool in place for several new commercial ventures," the *Business Journal* said, noting Titan's satellite scrambling system, medical instrument sterilization, and a pay-per-use data retrieval system.[34]

In hindsight it became crystal clear that Titan had succeeded in its commercial endeavors because it hired experts in the field, which included people who knew how to market commercial technology. "We went into commercial technology with our eyes wide open," said board member Daniel Fink, a former senior vice president at General Electric. "We knew that we had to avoid some of the pitfalls that the aerospace companies had fallen into when they tried to move into commercial applications. One of the reasons the aerospace companies were failing was because they didn't have marketing experience when they went into a commercial area. We at Titan brought in outsiders who essentially came from the world we were entering, and that was probably the single most important decision that enabled Titan to be successful in many of those areas."[35]

"We made conscious decisions," added McDougall. "When we started something that was commercially oriented, we took it out of the DoD management chain and gave it its own little entity inside of the company to ensure that it didn't get bogged down by all of the DoD requirements."[36]

Titan's leaders may have had the vision, but Ray and Webb credited Titan's employees with carrying the company through difficult times. "Our market strategy recognizes the importance of the hard work and dedication of Titan employees to the strength and success of the company," they

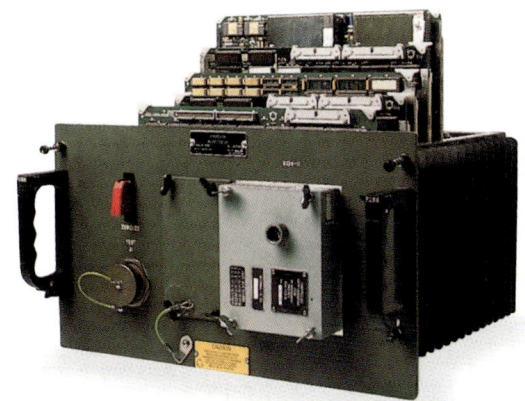

In support of the Milstar satellite communications program, Titan provided systems engineering and integration and developed terminal equipment.

emphasized in the 1992 annual report. "Drawing on that strength, we are convinced that Titan will not only cope successfully in a rapidly changing business environment, but that the company will help lead the nation toward its goal of greater global competitiveness and a full realization of the benefits the information age has to offer. We fully expect Titan to grow and flourish into the next century."[37]

Titan's satellite-delivered communications services provide voice and data links that are not geographically or economically feasible with terrestrial lines.

CHAPTER SIX

INNOVATION AND ENTREPRENEURSHIP
1993–1995

He's a risk-taker, always has been. Some people would have trouble operating the way he does. But look where it's gotten him.

—Kathy Pratt, discussing her father, Gene Ray

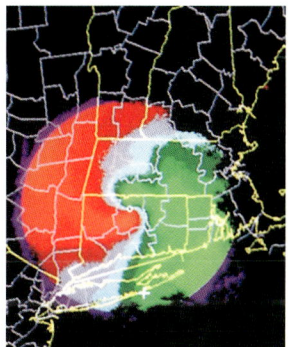

B Y 1993 IT WAS CLEAR THAT Gene Ray's willingness to try new ideas was a key ingredient in Titan's successful diversification. Ray was an entrepreneur and welcomed that spirit in others. He nurtured creative thinking and continued to draw new technologies to Titan.

"Recognizing opportunities and then being willing to take the associated risks is part of what has made Titan so successful," said Ken Kreyenhagen, who retired in 1995 as vice president in charge of Research & Technology. "Titan's success has come from that willingness to take risks and the recognition that you're not going to hit a home run every time."[1]

Caught in a Stranglehold

One such disappointment occurred early in 1993.

In June 1992, Titan had announced with much fanfare the formation of Titan Satellite Systems Corporation, a joint venture with Houston Satellite Systems. This new entity had exclusive sales and marketing rights for the Linkabit smart card, a satellite television scrambling system designed for secure, reliable, and cost-effective distribution of high-quality video and stereo audio. Put simply, the smart card system scrambled and unscrambled pay-per-view and other pay-television signals, ensuring that only those who paid for programming could view it.[2]

It was a concept with great potential, but there was a drawback: General Instruments (GI) had held a monopoly in the industry in the United States for some time. For the time being, GI's control would prove too much to overcome in the U.S. market.

After sinking more than $2.5 million into the technology, Titan closed the doors on Titan Satellite Systems and laid off 15 workers in February 1993.[3]

Still, Titan won kudos from industry experts for trying the new venture. "The trouble with this project really relates to the difficulties of trying to break into a monopolistic industry," said Robert Gutenstein, director of research for the investment firm Kalb, Voorhis & Company in New York. "It was a real long shot. It was a great try. The kind of return they would have gotten [if successful] was huge."[4]

"There was widespread hope in the industry that Titan would succeed," said Bob Scherman, editor and publisher of *Satellite Business News.*

Titan continued running tests and evaluation for the FAA's NEXRAD Doppler radar system, which could detect hazardous wind shear and weather patterns over airports.

Titan was viewed "as the first legitimate competitor to General Instruments' stranglehold on the marketplace. And I think there were a lot of people in the industry that were disappointed when Titan Satellite Systems failed."[5]

Mini-DAMA/Major Frustration

The satellite scrambling frustration was not the only problem that Titan would face in 1993. The company and the U.S. Navy became locked in a dispute over the funding of a major Linkabit communications contract for Mini-DAMA, a state-of-the-art satellite communications terminal whose name reflected significant reductions in both the size and the weight of the regular DAMA terminal. The dispute pushed Titan into a legal battle with former Linkabit owner M/A-COM.

The wrangle led not only to court but to a major cash crunch for Titan as the company continued to pour money into the project, pay court costs, and invest in new technologies. The crunch caused Titan to default on bank agreements and led to an $8.6 million loss in 1993, its first loss in five years.[6]

Fortunately, Titan was able to resolve the dispute by early 1994 with a $10 million settlement from the Navy. In the company's annual report, Ray focused on the positive. "First, the settlement helped to solve cash flow requirements," he told shareholders. "Second, the Mini-DAMA program was Titan's last large, high-risk, fixed-price development contract. This successful resolution significantly lowers the potential for major contract losses on existing and future defense contracts."[7]

Ray added that the Mini-DAMA had performed so effectively that it had been deployed early on aircraft destined for Bosnia, and its capabilities had piqued interest from international customers. By 2002, the Mini-DAMA program had accrued more than $230 million in revenue for Titan.

Partnering Up

Though Titan faced multiple challenges in the early to mid-1990s, it continued its strategy of joining forces with other companies to increase business.

In April 1993, Titan and Patriotic Scientific Corporation formed a strategic alliance to market Patriot's ground-penetrating radar technology coupled with Titan's electronic systems capabilities. The companies planned to market the technology for both government and military applications.[8]

In September 1993 Titan formed a strategic alliance with Integrated Cargo Management Systems (ICMS) to apply Titan's PositCOMM mobile communications technology to ICMS's ProfitMAX intermodal cargo container tracking and monitoring system. Titan granted ICMS an exclusive license to market its existing PositCOMM mobile communication and automatic vehicle location technology to certain segments of the transportation industry on a worldwide basis. In return, Titan was to adapt PositCOMM to ProfitMAX and manufacture the units for ICMS.

Then in December 1993, Titan signed a teaming agreement with Message Processing International (MPI) to create and implement programs for disaster preparedness, response, and recovery.

Titan was already working on several pro-

Titan's Mini-DAMA satellite communication terminal is deployed on Navy E-2C aircraft and are used to control carrier and land-based fighter aircraft.

grams in crisis management and emergency operations, and the two companies planned to work together on various projects, including use of MPI's NovAlert emergency notification system for the live implementation of Titan's simulations and tabletop exercises.

Titan formed yet another business alliance in September 1994 with Delta Data Systems. Together, the two companies won a $5 million contract to provide NATO members with a secure communications and information-handling system. Delta provided overall programs, management, PCs, and hardware. Titan provided software, system integration, documentation, installation, and training.

Titan also continued its agreement with Motorola in 1994 to upgrade Motorola's LST-5 series UHF Manpack radios with DAMA capability.

Strong Defense

In 1993, the U.S. defense budget dropped for the eighth consecutive year, a decline that did not go unnoticed at Titan. In April 1994, the company sold its Army training simulations business to San Diego's Cubic Corporation for $21 million in cash.[9] The sale not only supported Titan's move into information systems and commercial ventures but also helped pay off bank debt. In the 1994 annual report, Ray and Webb wrote that the divestiture would allow the company to focus financial, technical, and management resources on two business segments: information systems and applied technologies.

Yet Titan's defense contracting was far from over. The company continued to win military contracts, including a three-year, $4.9 million contract from the U.S. Air Force to produce and install a klystron test facility at McClellan Air Force Base in San Carlos, California. The new facility would test the E-3 klystrons used in the Airborne Warning and Control System (AWACS) radar. Klystrons generate radar transmission signals that are a crucial part of the United States' early warning system. Other major defense work came in early 1995, when Titan Linkabit won a $12 million contract to supply the Navy with Mini-DAMA. In addition to the initial production order, Titan provided options worth approximately $70 million that the Navy could exercise over two years.[10]

A Model for Success

Titan's strategy was to continue to build defense businesses and use the technology and cash they generated to build commercial businesses. "We expect our nondefense business will continue to grow," Ray told employees and stockholders at the company's annual shareholders meeting in May 1993. He said that about 500 of Titan's 1,325 employees worked at least part-time on some of the more than 20 commercial projects under way at the company. Asked to quantify how much commercial business Titan hoped to capture, Ray said simply, "As much as we can get."[11]

In the late 1980s, Titan had identified 54 different technologies, products, and concepts that had potential for commercialization. That list was narrowed down in the early 1990s to 14, each of which fit into one of four Titan commercial businesses: (1) broadband (video encryption); (2) satellite communications; (3) commercial information technology; and (4) medical product sterilization.

Titan produced and installed a test facility at McClellan Air Force Base to test E-3 klystrons used in AWACS radar.

Ray was prepared to back up Titan's commercial efforts with the necessary research and development, announcing that the company's ongoing commitment to an aggressive level of investment in the four businesses would allow it to introduce a portfolio of products and services that would maximize value in the long term.[12]

At a presentation before the Los Angeles Society of Financial Analysts in September 1994, Ray laid out Titan's long-term goal: "Our mission is to build a premier, technology-based, customer-focused information systems and services company . . . known for innovatively exploiting its intellectual property and technology base."

Titan's strategy was becoming known. In June 1993, Ray was invited by the Computer Electronics and Marketing Association to speak on how defense companies or other firms considering expansion alternatives could establish themselves in commercial markets. Then in April 1994, the *San Diego Union-Tribune* published a telling article about the state of local defense companies, noting that "in a region still grappling with the effects of defense cuts, Titan Corp. may yet emerge as a model defense conversion success story."[13]

One reason Titan succeeded in commercial endeavors where other defense-based companies failed was crackerjack adaptability. As Titan founder Jack McDougall pointed out years later, "A lot of defense contractors stumbled and failed at making any kind of conversion to the commercial side because the Department of Defense has such structured methods, and the development cycle for commercial products is a lot shorter than the DoD's used to be. Titan, however, was very good at overcoming those hurdles."[14]

Titan Information Systems

In yet another commercial move, Titan turned its information systems division into a wholly owned subsidiary in February 1994. Frederick Judge, former senior vice president and chief operating officer of Hughes Communications, was named president and CEO of San Diego–based Titan Information Systems.

Titan's Mini-DAMA terminals (inset) were installed in a number of military aircraft, including the E-2C Hawkeye (above). The terminal linked the aircraft to a global satellite communications system.

Titan helped the U.S. government push itself into the information age. The company's efficient information management systems supported the secure handling of data, which is essential for civil government and Department of Defense customers.

"We expect the information market to be among the most active segments of the economy in the 1990s," said Judge. "The growing importance of computers and high-speed telecommunications will increase the dependence on information products and professional information services such as those offered by Titan, for at least the next decade. We will be ready to meet those demands."[15]

Judge planned to advance three innovative businesses, including Titan's video encryption technology, which the company renamed Video PassPort™. "What we've found is that in this country, there's virtually no chance that we're going to penetrate [the GI monopoly]," said Judge. "But the good news is that around the world, there is starting to be an explosion of demand for these kinds of systems."[16]

Overall, Titan Information Systems would operate in three areas: secure television, which included Video PassPort; satellite communications; and custom software development and information technology solutions.[17]

In the area of satellite communications, Titan Information Systems' principal products included a network management system for efficient use of satellite bandwidth, compression products, quick-acquisition communication modem products, and rural telephony systems. Titan's network management product for DAMA satellite communications was called DAMALink. It provided voice, fax, data, and video access for multiple users on a shared basis.[18]

In software development, Titan Information Systems offered clients customization and re-engineering of operational support systems to stay on top of changes in the marketplace. The company also provided complete systems for storage management of massive quantities of data.

The Bumpy Road to Achievement

Analysts who followed Titan's steps into the commercial arena in the mid-1990s were on

Titan's innovative Video PassPort™ system garnered its first contract in 1995. Soon customers in New York City were using its equipment and box decoders.

the fence as to whether the new ventures would succeed. "Pursuing commercial business from within the defense framework has not been very successful historically . . . so this is a new model," said Jon Kutler, president of Quarterdeck Investment Partners.[19]

"I don't want to throw a brick through their window, but I don't want to throw roses either," said Robert Gutenstein of Kalb, Voorhis. "They're using the right strategy by going out to get expert people in markets that they've never competed in, and that improves the chances of success. But the ultimate measure is, will they come up with products that sell?" He claimed that plans for Titan Information Systems were "mostly gonnas right now. They're gonna do this and gonna do that. They are good guys, trying hard, but it's a tough road."[20]

Ray was quick to respond to Gutenstein's comments. "When you go from $300,000 in sales in 1990, to $3 million, then to $20 million in 1993, all in commercial information services, that's more than gonnas," he said. "Every road has potholes, and you don't usually see them. You just feel them when you hit them. So I'm sure we'll find potholes, but I don't see cliffs."[21]

Years later, in 2001, Ray would note that Titan's commercial business had helped grow the company nearly 100 percent in the short span of five years. "At the time, we got hammered like mad by the stock market because we were losing money. But that money we were losing wasn't in defense. We were losing money because we were making investments in these commercial endeavors that are paying off. If we hadn't stuck to it, we would not have a company today."[22]

More Talent

Titan continued its practice of seeking out, hiring, and promoting top managerial talent.

In 1993, Stephen P. Meyer, CFO, was promoted to president of Titan's Applied Technology Group, which included medical product sterilization. Meyer was responsible for getting Titan's first medical product facility, in Denver, operational.

Ronald B. Gorda was elected a Titan vice president in May 1994. Gorda, president and general manager of Titan Linkabit, was formerly senior program manager for Rockwell International, where he directed satellite communication programs for the Command and Control division.

C. L. (Neil) Hensel became senior vice president of Titan and president and general manager of Titan Systems in January 1995. Hensel had served as senior vice president and general manager of the C^3I Systems division of the Atlantic Research Corporation Professional Services Group.

David A. Hahn joined Titan in May 1995 as senior vice president, general counsel, and secretary. Prior to joining Titan, Hahn had been a partner at Latham & Watkins, where he maintained a general transactional practice with an emphasis on corporate and securities law.

Ken Kreyenhagen, the founder of Titan Research & Technology, brought in A. Anton "Tony" Frederickson, who had worked in Titan Research & Technology before taking a position in the DoD, to take his place when he retired in 1995. Frederickson had been chief of the Special Projects division for the Defense Nuclear Agency.

Finally, two of Titan's cofounders left Titan to pursue new companies—Jack McDougall in 1993 and Ed Knauf in 1995. Both men had held the title of executive vice president, and both had served on Titan's board of directors.

Restructuring

Early in 1995, Ray and Webb sought the board of directors' help with a particularly rankling challenge. They saw the potential for tremendous growth and profit in some of Titan's commercial ventures, but they needed to invest capital to produce that growth. Up until then, Titan had used its own financial resources and a back line of credit to fund such ventures, but cash was running low.

With the board's support, Ray and Webb determined a new strategy. "Our plan is to eventually spin out a minority interest in one or more of our businesses through public or private offerings to secure the capital necessary to continue to grow these businesses," they wrote in the 1995 annual report. It was a strategy that would prove successful in future years, but it took time to implement.

To better position the company for growth and strategic transactions, they restructured Titan into four business segments: Communications Systems, Software Systems, Defense Systems, and Emerging Technologies.

"We believe the segment reporting will enhance the investment community's understanding of our businesses and its ability to evaluate them," they wrote.

Under the new system of reporting, it was clear that Titan's Software Systems, Defense Systems, and Emerging Technologies segments had been profitable for the year. The Communications Systems segment did not show a profit, not because it was faring poorly, but because it was a business characterized by large projects, many of them with long lead times. Moreover, the projects were in emerging international markets and had all the attendant delays.

The restructuring also involved a number of changes in Software Systems that were designed to enhance its revenue and profit potential. The mass storage business was closed, and a new unit was created to focus on customers who were changing their methods of doing business.

The Evolution Continues

By early 1995, the *San Diego Union-Tribune* was characterizing Pentagon spending cuts as "a train wreck in slow motion." UCLA economic forecasts for three sectors of defense and aerospace businesses in California warned of a downward trend continuing for at least two more years.[23]

San Diego's General Dynamics, once a local defense giant, had moved positions, eliminated others, and was on the verge of shutting down its once huge Convair division. Tom K. Lieser, a research economist for the UCLA Business Forecasting Project, told the *Union-Tribune* that defense firms, "like anyone that's having to scrap for their livelihood, are experiencing mergers and consolidations and doing whatever they can to cut costs and increase efficiencies."[24]

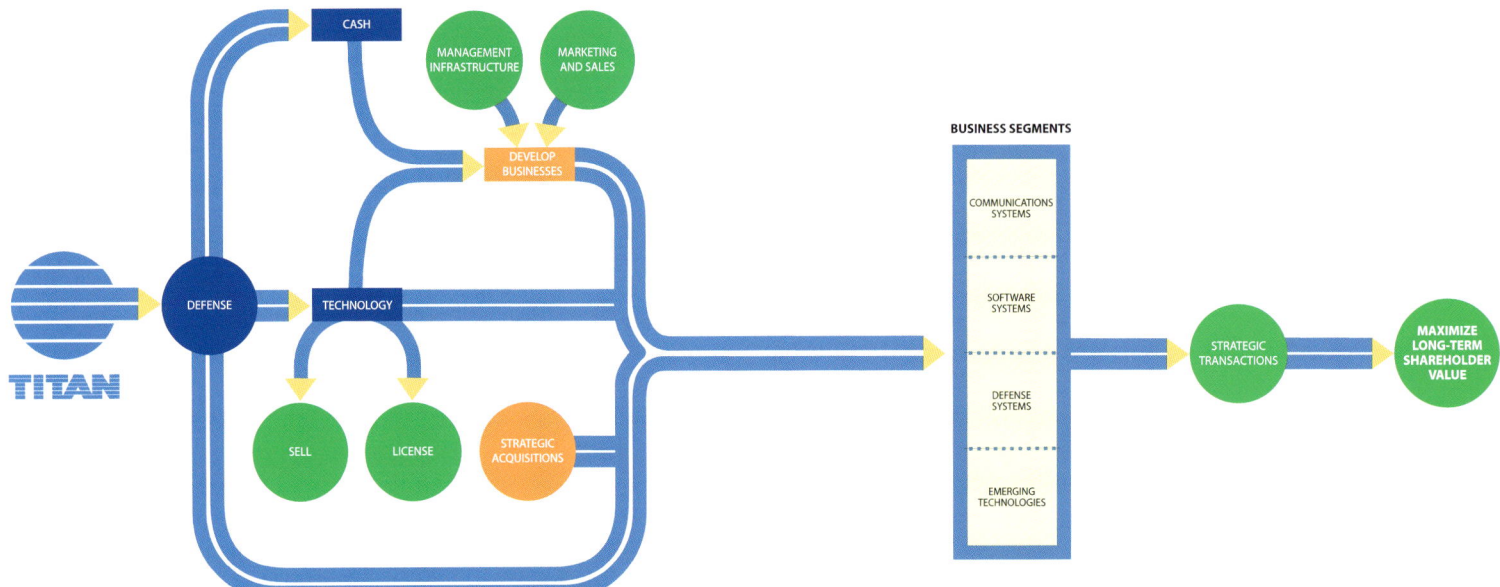

This flowchart represented Titan's new strategy to "create, build, and launch technology-based businesses."

Ray was not optimistic about defense prospects. "I don't think we've hit the bottom yet," he said, adding that even if defense spending were to increase immediately, "The pipeline has just been so cleaned out that it's going to take awhile to fill it up again."[25]

In January 1995, CellularVision Technology & Telecommunications (CT&T) awarded Titan Information Systems its first contract to supply Video PassPort™. CT&T used the system to control access to its patented wireless broadband interactive information system, which allowed operators to deliver multichannel interactive television to subscribers in metropolitan areas. The first Video PassPort system was installed in New York City.[26]

Then in April 1995, Titan Information Systems received a $2 million contract from Loxley Public Company in Bangkok to provide a satellite-based voice, data, and videoconferencing network for use by the Siam Bank of Thailand. And in July 1995, Linkabit won a contract from Motorola to produce satellite communications modems to be installed in Motorola LST-5 lightweight tactical radios.

A $10 million contract from Pasifik Satelit Nusantara in September 1995 further broadened Titan's international footprint. Under the agreement, Titan would develop, manufacture, and deliver satellite communication terminals, a network management system, and telephony access equipment for voice, data, and fax services throughout Indonesia.

The company continued to invest in Titan Scan. Titan opened its second electron-beam sterilization facility in November 1995 in San Diego to service a five-year, $5 million contract from IMED Corporation. It was the first such sterilization facility in the San Diego area, and Titan Scan managers hoped to attract local business.

By then, the SureBeam® process was already being used in Europe. BSE-Mediscan in Kremsmunster, Austria, bought a $5 million SureBeam unit in June 1995 for sterilizing medical equipment from various European facilities.

New government and military-related contracts were still crossing Titan's transom. In June 1995, Titan received a $5.7 million subcontract from Tracor Aerospace in support of the Explosive Standoff Minefield Breacher Demonstration and Validation program, sponsored by the U.S. Army and Marine Corps. The purpose of the program was to develop the capability to neutralize mines from a safe distance.

Takeover Attempt

Titan was in court in November 1995, attempting to thwart a potential hostile takeover by two of its biggest shareholders. Titan executives accused Robert K. Pryt, of BKP Capital Management, and Robert J. Feibusch, of Feibusch & Company, who together held a 12.9 percent stake in Titan, of flouting securities laws by spreading false information about Titan and trying to manipulate Titan stock.

Pryt and Feibusch had filed documents with the Securities and Exchange Commission calling for Titan to focus its interests and sell unrelated business units. They recommended that the company close down the medical sterilization business, Titan Research & Technology, and an environmental consulting and engineering arm of the company. Concerning Titan Scan, Feibusch wrote that it was "indefensible for Titan to choose to spend over $10 million of precious capital to enter this business."[27] They also indicated that they would run their own slate of directors.

Ray insisted that Titan would not depart from its corporate strategy. "What they want us to do is change the strategic direction of the company and sell off pieces," Ray said. "We have a strategic plan that we outlined three years ago, and we have been following it and it has been working."[28]

By March 1996, the two sides had reached an agreement. Titan would drop the suit, and Pryt and Feibusch would stop soliciting proxies or consents regarding Titan securities. Interestingly, BKP Capital Management went bankrupt a few years later in one of the nation's longest-running bear markets.

Titan discovered new uses for its electron beam technology, adding electronic irradiation of food products to sterilization of medical equipment.

CHAPTER SEVEN

CREATING VALUE

1996–1997

The company needed cash, and it had a limited number of options. We drew the line in the sand in April 1997 when we realigned the company into four different businesses.

—Eric DeMarco, 2001

THOUGH TITAN WAS ABLE TO resolve its legal battle with Pryt and Feibusch, the conflict raised disturbing questions on Wall Street. Were Titan executives confused about where the company was heading, as the disgruntled shareholders had complained? Was Titan a defense company, a software company, or a medical sterilization company?

These questions, combined with the fact that Titan was still investing in commercial businesses and experiencing a cash crunch, led its stock price to drop nearly 50 percent in 1996. Ray attributed the drop to the company's continued investment in commercial businesses and to being unable to communicate to Wall Street the future payoff that would come from those investments. He told the *San Diego Union-Tribune* in September 1996 that Titan was actually on target in its mission to remain in the defense industry while exploring and developing commercial opportunities.[1]

That year, the Titan strategy began to materialize for its shareholders. One commercial business, television encryption, was discontinued and sold, and the other commercial businesses began to mature. Wall Street finally started to understand the payoff, and Titan's stock price went up accordingly, allowing for three successful commercial businesses.

Going Strong

In the meantime, the military communications systems segment continued to perform well, and doubts on Wall Street regarding Titan's ongoing work in military communications began to fade. As a 1996 study by the strategic marketing firm of Frost & Sullivan noted, "The military communications market is very much alive. Despite the changing military environment and downward pressure on defense spending, communications will remain a vital element of any conceivable force structure."[2] Titan Linkabit, under Ron Gorda's able management, proved this to be true for Titan.

Early in 1996, Titan won a $2.6 million contract to upgrade some of the U.S. Army's satellite terminals to make them capable of accessing commercial satellite bandwidths. The contract, awarded by the Army's Communications and Electronics Command, called for Titan to provide the Army's LST-8000 terminals with tri-band capability. The terminal upgrade aptly demonstrated Titan's

---

Titan did a great deal of software work for the telecommunications industry, where deregulation created intense competition. Titan's software offered an advantage to users by decreasing the response time to new information needs.

The U.S. Navy continued to be one of Titan's largest customers. Titan's work for the Navy included engineering support and systems installation, integration, and maintenance.

success in expanding military technology to commercial communication networks.[3]

In March 1996, Titan won a $2.3 million Milstar contract to provide engineering support for on-orbit testing of UHF Milstar and to upgrade the dual-modem satellite communications modem for the U.S. Air Force. Two orbiting Milstar satellites underwent extensive engineering and operational tests with the dual modem. The technology of the upgrade was a cost-effective means of allowing existing systems to operate with the next generation of military satellites.[4]

That summer, Titan won $2.8 million worth of new contracts to provide engineering support and installation of communication systems for the U.S. Navy. Under these awards, Titan would install meteorological, oceanographic, and communication systems on Navy ships in Japan.

A few months later, Titan received a $2.5 million order for LSM-1000 UHF DAMA satellite communications modems from the Navy and DynCorp Aerospace Technology. This was the first large contract for LSM-1000 modems produced by the Linkabit division.

Global Expansion

Titan also continued to build business internationally. The company initiated the production phase of its contract with Indonesia's Pasifik Satelit Nusantara (PSN) during 1996 by delivering more than 100 satellite-based, proprietary Xpress® Connection remote terminals. Xpress Connection used existing satellites to provide low-cost voice, fax, and data services to areas of the world that did not have access to phone service. In addition, PSN announced plans to install a significant number of Titan's terminals throughout Indonesia to reach more than half of the 30,000 villages that had no telephone service.

Mike Kulinski, senior vice president for system development and Titan's chief technology officer, explained that when dealing with remote areas or single villages, it was easier to provide wireless, satellite communications than to install fixed lines. "But you still have a village where you need to install this infrastructure," he said. "Indonesia is a very diverse country. It's the fourth largest in population, with over 200 million people on thousands of islands. We connected those islands with VSAT [very small aperture terminals]," which, he noted, provided voice, data, and Internet access. "A lot of farming communities cannot get crop updates or crop pricing," he said. "So our technology can actually help them improve the quality of life as well as maintain communications."5

Having an affordable price on the technology was a key to winning the contract in Indonesia, as Kulinski explained. "We had to come up with a very low cost, extremely efficient terminal for these marketplaces. When we started looking into it, the benchmark for satellite terminals was anywhere between $10,000 and $20,000 a terminal." But, Kulinski explained, by applying Titan's DAMA technology and other high-performance modem technology that Titan had invested in, the company was able to produce terminals that sold for only $3,500.6 This was not the first time Titan had applied its existing technology to other products, and it would not be the last.

The Indonesian installation sparked interest in Thailand. In July 1996, Thailand's United Communication Industry Public Company (UCOM) awarded Titan a $3.97 million contract to provide voice, data, and fax services to the Bank for Agriculture and Agricultural Cooperatives.

"The government of Thailand does not want the farmer to move from the farm to Bangkok, which is very overcrowded," explained Kulinski. "They want to keep their agricultural business as successful as possible. So we've provided these small, rural banks with communications and data services by installing a small terminal there. And now the farmer, if he needs a loan to buy seed, for example, can use the wireless network to communicate from the farthest regions of Thailand directly to Bangkok. His loan can get processed almost overnight, and then his small bank branch can distribute funds."7

Pursuing such international contracts was a sign of Titan's commitment to providing satellite communications solutions to developing nations. In some of the most rural, least developed areas of the world, Titan's wireless technology helped families and businesses communicate better. This interchange was especially meaningful considering the remoteness of many countries' villages from their major cities, where most businesses were located or where the family's wage earner often worked and lived. Medical emergencies, too, could be better handled in remote areas when residents could communicate with cities that had comprehensive medical service.

Beyond the social implications, Titan saw a profitable future in providing wireless information and telephony services to developing countries, where local governments couldn't afford to lay land lines for even the most basic telephone service.8

In the international military arena, Titan was awarded a $2.9 million contract in September 1996 by the NATO Consultation, Command and Control Agency in Belgium. Under the agreement, Titan provided TEMPEST hardware and software products, as well as system integration, documentation, installation, and training at NATO military headquarters.9

More Commercial Highlights

Titan Information Systems got a big break for its Video PassPort technology in 1996 with the award of a $12.75 million contract from Cellular-Vision Technology & Telecommunications. Under the agreement, Titan delivered conditional access system equipment and set-top box decoders for customers throughout the New York City area, allowing multichannel television programming and interactive entertainment. At that point, Titan was still testing the

Titan's satellite communications systems allowed people in developing nations and remote areas to enjoy telephone and other telecommunications services for the first time. In 1996, Titan began installing satellite terminals in Indonesia.

In 1996 Titan acquired three defense technology companies, all involved in integration, installation, and maintenance of information systems on U.S. Navy ships and submarines.

waters to see if this would be a truly profitable venture.

Titan Software Systems suffered a decline in revenue in early 1996 but continued to grow according to plan. Software for telecommunications was a particularly strong area. After passage of the 1996 Telecommunications Act, which deregulated the industry to promote competition and better quality, a number of long-distance suppliers, local exchange carriers, and wireless carriers looked to Titan's software solutions to help them become more competitive. Titan's telecommunications customers included MCI, GTE, Nynex, and Bell Atlantic. The software segment was also at work on contracts from the FAA and commercial banks.

In the company's Emerging Technologies segment, Titan won a $2.4 million contract from the Taiwan government to produce MULTIBUS and VME computer products. The equipment was used in data processing, communications, fire control systems, and early warning systems.

Defense Systems

In keeping with Titan's mission to focus on its core businesses of communications and information systems and services, Ray announced in the summer of 1996 the sale of the company's electronics division to SCI Systems, an aerospace and government electronics manufacturer.[10]

While divesting in one area, Titan was making strategic acquisitions in another. In May 1996, the company announced the purchase of three affiliated defense technology businesses: Eldyne, of San Diego; Unidyne, of Norfolk, Virginia; and Diversified Controls Systems, of Richmond, Virginia. The three companies, purchased for $23.6 million in stock, cash, and other considerations, were involved in integration, installation, and maintenance of information systems on U.S. Navy ships and submarines.

Bud Leedom, publisher of the *San Diego Stock Report*, told the *Union-Tribune* that the acquisitions seemed to run counter to the company's strategy of diversifying into commercial areas and that it was difficult to tell where Titan was heading.[11] However, Titan had never downplayed defense as a business, and Ray predicted that adding these companies would strengthen Titan's existing Defense Systems segment, increasing its capability to win contracts. In addition, the acquisition generated cash to support commercial ventures and to provide new technologies with the potential for commercialization.

It wasn't long before Ray's predictions came true. In July 1996, Titan received additional orders from the U.S. Navy and an independent defense systems contractor for an airborne version of Mini-DAMA, to be produced by the Linkabit division.

An $8.2 million production option was added in October 1997.

In October 1996, Titan won a $1.6 million contract to test and evaluate tactical data dissemination and related C³I systems for a DoD contractor. Options would increase the value of the contract to $7 million.

That same month, Titan won a $1.5 million agreement from Comarco to provide engineering and technical support in the development of Test Program Sets for the avionics systems on B-52 strategic aircraft.

Most impressive of all, Titan's Unidyne subsidiary, under the leadership of Dave Conner, kicked off 1997 with a five-year, $25 million agreement with the U.S. Navy to provide engineering and other support services to the Atlantic-based Fleet Technical Support Center. The services bolstered submarine undersea warfare and other electronic systems.[12]

A Capital Idea

Despite Titan's recent successes, Ray knew he needed a financial boost to keep the company on track. Titan was making progress in its commercial ventures, but investments in those ventures had also caused an ongoing cash crunch, which amounted to a $3.8 million loss for 1996. Titan needed more capital—and the sooner the better.

In the fall of 1996, Ray announced plans to raise $30 million by issuing bonds that could be converted to stock at a later date. Titan told the Securities and Exchange Commission that proceeds would go toward paying off $18.6 million in short-term debt, and the remainder would be used for working capital, joint ventures, and acquisitions.[13]

It was a short-term solution, but Ray was also appraising the long term. He determined that Titan needed a financial wizard who could work with him

In 1996, Titan began providing engineering and technical support for avionics systems on B-52s.

in creating a master plan for the company. As CEO, Ray had enormous vision and was superb at recruiting talent, but money decisions were not his strongest suit.

Roger Hay resigned as Titan's chief financial officer in the fall of 1996. He was succeeded by the talented Eric DeMarco, whom Ray brought on board in January 1997.

DeMarco had been a senior audit manager with Arthur Andersen & Company and had handled Titan's account for Andersen since 1985. He joined Titan as CFO and senior vice president.

Analysts heartily approved of the hiring. "Now you've got a real number cruncher with your visionary," said David Weinstein of Joseph Charles & Associates. "That is your secret formula."[14]

It was a formula that served Titan well financially and in terms of attracting other new talent. M. C. "Bud" Baird, whose private company, Delfin Systems, would be acquired by Titan in 1998, said that DeMarco was one of the reasons he agreed to join Titan. (Baird later became president and CEO of Titan Systems Corporation.) "I was very impressed with Eric," Baird said. "I felt that he was exceptionally knowledgeable and capable and that he would be a major element of Titan's success."[15]

Ray and DeMarco wasted no time putting their "secret formula" to work. Titan needed to find a way to financially reward the top-notch commercial talent it was recruiting. "So we visualized and conceptualized the holding company structure," DeMarco said. "That way we could bring in management with its own stock options. The management team would have a liquidity event possibility through either an IPO or a sale."[16]

At the same time, Titan began transitioning to a new, longer-term business strategy. "On the macro level across the company, we tried to get away from selling a product and shifted instead to providing a service and an annuity," DeMarco said. Rather than trying to make multiple one-shot sales, the company would broaden its business by servicing customers over the long term.[17]

Spin-Offs and Spin-Outs

Early in 1997, Ray and DeMarco began putting their joint vision to work by realigning the company into four focused companies: Titan Technologies and Information Systems (defense); Linkabit Wireless (defense and commercial communications); Titan Software Systems; and Titan Sterilization and Pasteurization Systems, which included a segment that focused on emerging technologies. Titan planned to profitably grow each business and increase operating efficiencies while maximizing shareholder value through strategic transactions.

Ray and DeMarco's idea for "strategic transactions" involved spin-outs and spin-offs, which would prevent draining precious resources and draw in more capital. Whereas previously Titan had relied on its defense business to generate cash for commercial endeavors, now the company would seek outside capital through spin-outs of minority interests in those businesses that had accumulated sig-

Titan was well positioned in the exploding Information Age. Its software systems subsidiary specialized in integrating software solutions for commercial and nondefense government clients.

nificant shareholder value. The capital could come through an IPO or strategic partnering, for example.

Setting up employee stock option plans for potential spin-off and spin-out companies was key to the companies' success. As Ray often emphasized, finding the right people was critical, and employee stock option plans were prime incentives for recruiting and rewarding good people.[18]

The spin-out/spin-off design would determine Titan's future. Within the next few years, in fact, the powerful strategy would transition the company from a business leader into a business titan.

The Plan in Action

The spin-off of Titan's broadband communications business in the first quarter of 1997 illustrated the new strategy. Titan had successfully developed an analog and demonstrated a state-of-the-art Video PassPort conditional access system for television delivery systems. But things were not moving quickly enough. Titan was pouring a lot of money in, but PassPort wasn't generating enough revenue.

"The specific market opportunities for our broadband business have not developed as rapidly as was anticipated," Ray explained. "Given the capital demands of this business, we concluded that the best way to return value to shareholders would be through a spin-off or sale of the business."[19]

Another spin-off in the first quarter of 1997 further supported Titan's new financial strategy. Titan sold virtually all of the assets of its TEMPEST computing products division, Delta Data, to Cycomm International. This spin-off not only unburdened Titan but also gave the company future business. In sealing the deal, Titan and Cycomm agreed to a partnership in which Cycomm would be the preferred provider for Titan's secure computing requirements in systems integration projects. In turn, Cycomm would refer future systems integration opportunities to Titan.[20]

In May 1997, Titan established a new business called ServNOW! NetTechnologies to develop a priority software product. To get starting capital for the new business, Titan raised $4.2 million from a group of venture capitalists led by San Diego–based Enterprise Partners. Ray believed that the start-up—targeting Web-based content management and delivery for information providers, corporate intranets, and Internet service providers—could be more fully developed with outside capital.

"Given the rapid growth and growing complexity of Internet usage, ServNOW! is addressing a market opportunity with enormous potential," he said. "Having made a minimal initial investment in the business of about $200,000, Titan has determined the interests of our shareholders can be best served by introducing outside capital to fully exploit the potential of this technology. ServNOW! is representative of our strategy of leveraging our defense business technology to create new opportunities in the commercial sector."[21]

Ray may not have realized how on target he was. In just over two years, the business, renamed Ipivot, was sold to Intel for $500 million, and Titan realized $45 million for its remaining ownership. "The realization of the proceeds on Ipivot was a high point for us financially," said Virginia Oliver, assistant corporate controller who had been with Titan since the EMM merger in 1985. "Our proceeds from that were $45 million."[22]

Linkabit Wireless

With its four new business segments clearly defined, Titan began to make heady progress in 1997.

Linkabit Wireless flourished. The wholly owned subsidiary consolidated commercial and defense communications with an emphasis on rural telephony and secure defense communications. The subsidiary developed and produced satellite ground terminals, satellite voice/data modems, networking systems, and complementary products that incorporated DAMA technology.

Linkabit Wireless joined a consortium that was taking part in the Multi-Media Asia Satellite Telecommunications System (M^2A) project. The consortium, led by Alcatel Telspace, a French subsidiary of Alcatel, designed a satellite-based system that provided telephone, fax, Internet access, and television services to 4 million subscribers in Southeast Asia.[23]

Titan's Indonesian connections proved valuable to the consortium's success. In July 1997, PT Multi-Media Asia Indonesia, a joint venture between

Pasifik Satelit Nusantara and Indosat, chose the consortium to supply the ground segment for the M²A project, with Linkabit Wireless developing and manufacturing hardware and software for subscriber terminals. The overall value of the contract was $105 million.²⁴

Linkabit Wireless further grew its international business when it received a $4.5 million contract from Raytheon E-Systems for Mini-DAMA terminals to be used by the Royal Australian Air Force. Australia's AP-3C fleet conducted submarine and surface defense missions over major areas of the Pacific and Indian Oceans. Because most of these missions were conducted far from operating bases, the need for real-time, reliable satellite communications was critical.²⁵

Linkabit Wireless also captured some important repeat business. In July 1997, the company received a $2.8 million contract from Motorola for LST-5D satellite communication modems to be used in lightweight tactical radios. This marked the third order from Motorola for these modems and was indicative of the successful, close working relationship between the two companies.²⁶

By the end of 1997, Ray and DeMarco felt ready to spin-out Linkabit Wireless. Titan filed with the Securities and Exchange Commission for an IPO of 2.7 million shares of common stock with plans for Titan to own 70 percent of the company after the IPO. Business was booming, and Linkabit needed capital to expand further. The timing, however, turned out to be unfavorable. The Asian economy nose-dived in early 1998, and the IPO climate in the United States began to cool off. By the middle of 1998, Titan would be among a number of companies withdrawing their IPO bids.²⁷

Software Systems

Titan grew its software integration business through new sales and strategic alliances. In the summer of 1997, the company received a contract from the FAA Air Traffic Service for data integration and design and development of the first air traffic component of a customized intranet channel called Executive Information System. Under the agreement, Titan Software Systems developed and

Linkabit received a $4.5 million contract to supply Mini-DAMA terminals to the Royal Australian Air Force's AP-3Cs, similar to the U.S. Navy P-3, pictured. Submarine and surface defense missions also utilized the communication devices.

Titan Software Systems continued its work for the FAA. In 1997 it began developing a system to provide up-to-date information on daily and long-term FAA air traffic operations.

implemented a system that provided up-to-date information on daily FAA air traffic operations as well as long-term trend information. The system also provided decision-support information on operations counts, air traffic delays, operational errors, and operational performance.

Titan found it would best be able to serve the FAA by utilizing products from PointCast. Titan entered into a solutions partner agreement with PointCast that allowed Titan to deliver customized corporate broadcast solutions to the FAA using PointCast's Internet and intranet broadcast services.

It was another step forward for Titan in the area of corporate information management, data warehousing, and executive information systems. As Ray noted, "This marks our entry into a new market area. This is indicative of the areas in which we will be focusing our efforts as we continue to develop additional new technologies, products, and services for the Internet."[28]

Such partnering was a good way to increase business. In November 1997, Titan Software Systems entered into another solutions partner agreement, this time with Janus Technologies. This partnership allowed Titan to deliver extension and integration services to financial institutions that used Janus' Argis product set.

"As fast as technology is evolving, and as distributed computing environments are becoming increasingly complex to manage, companies will need new ways to meet changing needs," Ray said. "Our partnership with Janus provides a new way for customers to get this new challenge under control."[29]

In addition, Titan Software Systems worked with several large U.S. companies toward Y2K compliance.

Technologies and Information Systems

Titan continued to do well in its Technologies and Information Systems segment. Titan's defense revenue for 1997 increased 19 percent to $88.2 million. This accounted for more than half of Titan's overall revenue.

Despite all the consolidation going on in the defense industry, Titan continued to grow its defense segment internally and also sought out companies that would enhance its defense business. Over the next few years, that assertive, strategic acquisition strategy would pay off in a big way.

In the meantime, however, Titan Technologies and Information Systems continued bringing in contracts. In the first quarter of 1997, Titan received a $50 million task-ordering contract from the U.S. government. Under the five-year agreement, Titan provided engineering design, analysis, testing, and evaluation of various shipboard, vehicular, and shore-based radio communications and surveillance systems.

Another contract in 1997 was a five-year, $20.1 million contract from the U.S. Navy to provide engineering and technical support to the Naval Undersea Warfare Center Division for combat weapon systems. The Navy also awarded Titan a $14.1 million contract to provide engineering and technical support services to the DoD and other government agencies.

Sterilization and Pasteurization Systems

Titan continued to expand its medical sterilization business and found new uses for its SureBeam® technology during 1997. Early that year, the company discovered a new way to make profits with medical sterilization: by selling the technology and allowing other companies to use it on their own. This was part of Titan's overall initiative to "focus outward," in the words of Titan Vice President Tom Allen, one of Titan Scan's first employees. Allen noted that "instead of owning, operating, and developing products inward for ourselves, we began to look at the market, to add people and staff, and to look at the marketing tasks and functions. And we did a pretty good job."[30] Titan sold its first SureBeam turnkey system to Guidant Corporation.

Guidant was a leader in the medical device manufacturing industry and proved a valuable showcase to demonstrate the on-site system's numerous benefits. First, companies could make sterilization part of the manufacturing process, saving time and money involved in transferring materials to off-site sterilization contractors. Second, the system was small enough to fit into existing production facilities and portable enough to be relocated when production demands changed.[31]

Those benefits were enough to sell a second SureBeam unit to Baxter Healthcare in October 1997. The Baxter contract was valued at more than $4 million.

Titan was also investing in another way to use its patented sterilization process: for food, particularly ground beef. In 1993, an outbreak of *Escherichia coli* (*E. coli*) poisoning in the northwestern United States had been traced to the San Diego–based Jack in the Box's hamburgers. Four children had died, and 400 people became ill.

"That really changed the whole concern about the issues related to food safety," said Dr. Denny Olson, who led the food irradiation program at Iowa State University (the only food irradiation program at a U.S. university) for more than 15 years before joining SureBeam Corporation in 2000.[32]

SureBeam® offered a way to destroy the deadly bacteria and avert food-poisoning tragedies. As Titan's annual report explained, "SureBeam works much like a microwave oven. It takes ordinary commercial electricity as its power source and accelerates a stream of electrons into a powerful

---

Titan sold two turnkey SureBeam systems in 1997 that would be used to sterilize medical products. By the end of the year, the FDA had approved the use of electron beams to destroy bacteria in red meat, and Titan began aggressively marketing SureBeam for that purpose.

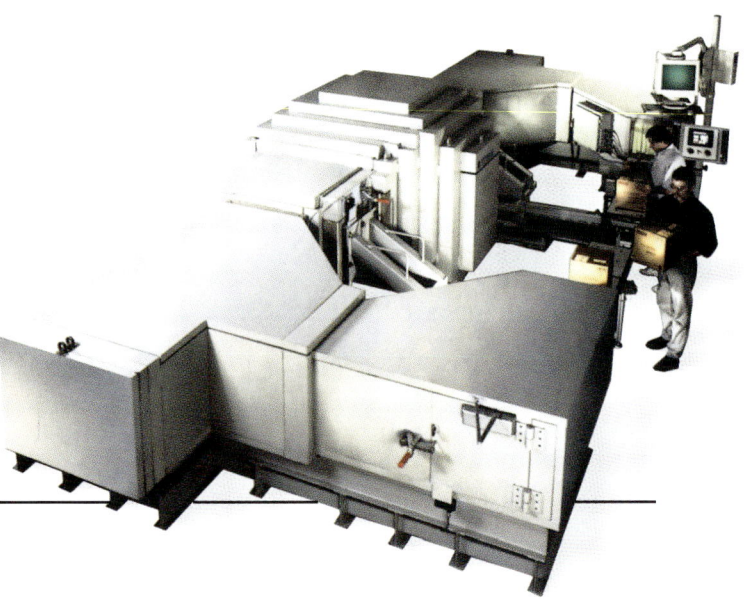

beam. When the beam scans food, it kills harmful bacteria in a flash—without changing the food's taste or texture."[33]

Brian Williams, who joined Titan in 1992 as technical director for Titan Scan and later became principal technical advisor to SureBeam Corporation, offered details of the SureBeam process:

The process reduces the pathogenic organisms to such a level that most people with generally healthy immune systems would not be affected by those microorganisms. When the food product is irradiated, the initial energetic electrons produce a shower of second electrons, and those are produced within the food itself. These secondary electrons then interact with the DNA of the microorganism and rupture the DNA chain. When that happens, the microorganism becomes incapable of reproduction and is essentially inactivated.[34]

The 1993 *E. coli* outbreak sparked major improvements in the U.S. government's meat inspection system and enhanced sanitation in meat plants, but those steps weren't enough. Millions of people every year were still becoming ill due to foodborne bacteria. It soon became clear that food irradiation was the best solution.

When the Food and Drug Administration in December 1997 authorized the use of electron beams to destroy harmful bacteria in red meat, it opened a door for Titan. "That was the major catalyst, what we would call the passing-of-the-red-meat petition," said Allen.[35] Still, it would be several more years before the United States Department of Agriculture (USDA) approved the same process. In the mean time, Titan continued to refine the life-saving technology and to lay the foundation for marketing it to the masses.

Profits Up

The new strategies Titan invested in during 1997 paid off rapidly and well. From a $3.8 million loss in 1996, the company made a turnaround to a $5.2 million profit in 1997. Total revenue was up nearly 27 percent, from $135 million in 1996 to a record $171 million in 1997.[36]

Chairman Sid Webb expressed his optimism for the future in the 1997 annual report, saying, "We believe the outlook for 1998 is promising. Our four core business segments are now well positioned to capture growth opportunities and pursue strategic transactions that we expect will finance and profitably grow our businesses."[37]

Titan Systems, the corporation's defense and information technology segment, continued to account for the bulk of Titan's sales, thanks in part to strategic acquisitions in that field.

CHAPTER EIGHT

MAXIMIZING VALUE

1998–1999

Spin-outs maximize the value of our business. And we know Wall Street loves pure plays. So we're going to give them pure plays with spin-outs.

—Gene Ray, 1999

ANYONE OBSERVING THE Titan Corporation early in 1998 could see that the investments over the past few years had positioned the company for real growth in commercial businesses such as sterilization and satellite communications. Gene Ray, Eric DeMarco, and the rest of Titan's leadership team had spent that time designing the launching pad for future success—and by 1998 Titan was in orbit.

It would be Titan's most spectacular year since its founding, with a record $300 million in revenue—an amazing 75 percent increase over 1997. Nearly every corner of the business showed improved operating performance. And a $620 million backlog of contracts positioned the company for sustained success.

Titan not only had a strong defense business still intact; it had successfully branched out into other areas, too. As analyst Thomas Meagher, of Ferris Baker, Watts, observed, "Titan is one of the few companies that has been able to successfully make the transition from a defense technology to a commercially viable technology."[1]

Ray and Sid Webb were happy to report in their 1998 letter to shareholders that Titan had "either met or exceeded consensus Wall Street estimates each of the last nine quarters." Throughout 1998 the company was able to stay on the success path with more spin-offs, more joint ventures, and a business buying spree. In March 1998, Titan once again demonstrated its skill at transferring a defense-related technology to commercial use with the spin-off of its TomoTherapeutics subsidiary.

TomoTherapeutics transformed a technology Titan had capitalized on in defense work into a cancer therapy technique. The X-Ray Needle enabled a medical device to emit energy from the tip of a needle so that the energy was deposited directly into a tumor, leaving surrounding healthy tissue untouched.

TomoTherapeutics licensed the X-Ray Needle to Influence Inc., a San Francisco–based medical device developer. Titan retained an equity stake in the company and received product royalties from Influence. Ray told *Medical Industry Today* that Titan's investment in developing the technology had been "minimal."[2]

Partnerships

Besides spinning off new and improved technology, Titan was able to grow the company through partnerships and alliances.

Titan was able to grow certain businesses best (and maximize shareholder value) through company spin-offs and spin-outs.

The company expanded its reach into the international telephony business with a new alliance in October 1998. Titan and Sakon, a two-year-old telecommunications company based in the United States, joined forces to provide international voice, data, fax, conference calling, prepaid calling, and other services to developing countries in Africa, Latin America, the Middle East, and Southeast Asia. Sakon was already providing telephony services to 10 countries in those areas. Its ability to deliver low-cost international minutes gave people in remote areas all over the world the ability to communicate beyond their immediate villages. Under their agreement, Titan obtained an initial 19 percent equity stake in Sakon and an option to increase its ownership to 50 percent.

Also in October, Titan bought 50 percent ownership in Afronetwork, an African telecommunications company, with the goal of building a satellite-based, national telephone system in Benin, Africa. The new system would be capable of providing telephone service to rural areas of Benin as well as interurban service with new wireless local loop capability.

Software Systems

Though Titan's software business had gone through some down times, it also improved its performance in 1998. A real feather in the division's cap came in September with a $34 million contract from the state of Wyoming for Y2K (year 2000) compliance services. Under the agreement, Titan, with Cap Gemini America LLC as subcontractor, provided an inventory and assessment of all of the state's information systems and remedied problem situations.

It was the largest contract ever for Titan Software Systems, and it was even more significant because not all states awarded their Y2K work to a single vendor. "For us, it's a significant win," said Tom Konchan, senior vice president and general manager for the software division. "It shows up on the radar screen."[3]

Sterilization and Pasteurization Systems

The turnkey SureBeam® medical sterilization system that Guidant Corporation had purchased from Titan Sterilization and Pasteurization Systems began operating in 1998, and Titan delivered another system that year, this one to Rochialle Corporation, in Wales. Titan also worked diligently on developing and protecting its food sterilization technology during 1998.

In addition, Titan Sterilization and Pasteurization Systems won several contracts during the year. The first was a five-year agreement from Valleylab, a division of United States Surgical Corporation, for the sterilization of disposable medical products. The second was a five-year agreement from American Precision Plastics to sterilize petri dishes, specimen collection containers, and laboratory testing plates. Finally, the subsidiary received a $2.7 million order for its linear electron accelerator, to be used for radiation chemistry research at the Commissariat à l'Energie Atomique in France.

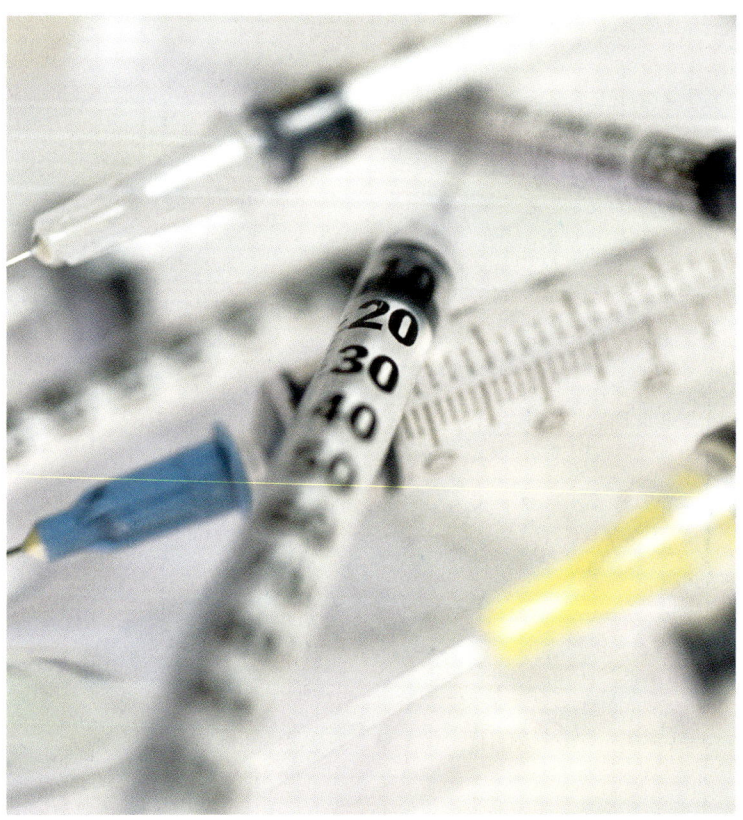

The SureBeam system was able to sterilize medical equipment after it was sealed in packages and boxes, thus eliminating the possibility of recontamination.

Shopping Spree

While Titan's existing businesses continued to flourish, Ray and DeMarco moved swiftly and boldly to expand through multiple acquisitions. The year 1998 turned into a veritable buying spree, with Titan acquiring five companies and positioning more for future purchase.

In March 1998, Titan purchased the publicly traded DBA Systems, based in Melbourne, Florida, in a tax-free exchange of stock. DBA had been founded in 1963 and developed and manufactured digital imaging products, electro-optical systems, and threat simulation/training systems. It primarily served the defense and intelligence communities but also sold products in other markets, including law enforcement and medical, transportation, and geographic information systems.

DBA became an operating company of Titan Technologies and Information Systems. With the addition of DBA, revenue for Titan's information systems subsidiary was expected to climb to $140 million annually.

As Ray had planned, this acquisition was quick to add value to Titan. The ink on the deal had barely dried when DBA received orders totaling $2.5 million for its ImagClear Model F5000 High Volume Fingerprint Card Scanner. These orders marked DBA's launch into the worldwide law enforcement fingerprint identification market. A similar contract for $2.2 million was awarded in October 1998, and another in May 1999. In addition, DBA won an $18 million operations and maintenance contract from the U.S. Army in May 1998 to provide software and hardware operations and maintenance support to systems developed by DBA at Army sites in Georgia, North Carolina, and Germany.

Next Titan acquired Validity Corporation, a privately held information technologies company based in Encinitas, California, for $15 million. Validity, with offices in Maryland and Arizona, had been founded in 1971 and supported both national defense programs and a variety of government agencies in the areas of systems engineering and integration, software engineering, network technologies, and test and evaluation services.

Titan expected Validity to generate about $29 million in revenue in 1998. Only days after the acquisition, Validity was awarded a $49.3 million task order to provide software development, maintenance, and products to the U.S. Customs Service.

In July, Titan bought one of its earliest direct competitors, privately held Horizons Technology, for $19 million in stock. Like DBA Systems, Horizons became part of Titan Technologies and Information Systems. Based in San Diego, with offices in Massachusetts and Melbourne, Florida, Horizons was particularly strong in the defense command and control (C^2) sector. It provided defense systems engineering and program management services, computer systems integration, and high-end software. Horizons had also developed a commercial software business, which included digital map information that could be purchased over the Internet.[4]

"We had a number of suitors, and we essentially picked Titan and Titan picked us," said Earl Pontius, Horizon's president and CEO. "It had to be a match that both companies were interested in and felt good about, and we really did." Pontius became president of Titan Systems' Technical Resources Sector and president of that sector's Information Solutions Group.[5]

The addition of Horizons promised to push Titan's government information technology business to revenue of nearly $200 million annually and would allow Titan to compete for larger contracts.

Also in July, Titan purchased VisiCom Laboratories, of San Diego, a hardware and software product innovator that specialized in communications, war game simulation equipment, image enhancement systems, and technical services in software and systems integration. Titan paid $25 million in stock in a tax-free exchange for the privately held company. VisiCom's annual revenue of $41 million was split equally between defense and commercial markets.

VisiCom was another good fit, and it quickly began bringing in profits for Titan. In the weeks following the acquisition announcement, VisiCom won $55 million in contracts with potential for $37 million more in follow-on orders.

The fifth acquisition of 1998 came when Titan purchased Delfin Systems, a privately held company in Santa Clara, California, in October. Delfin provided systems engineering and program management services, computer systems integration, and high-end software, primarily for the U.S. Navy and intelligence agencies. Founded in 1984, Delfin had particular strengths in intelligence analysis, computer forensics, information security, and enterprise communications and connectivity. The price tag was $22.5 million in stock.

M. C. "Bud" Baird, who had been president, chairman, and CEO of Delfin, became a Titan senior vice president and president of Titan Technologies and Information Systems. Baird agreed to pool Delfin's stock with Titan's because he was attracted to Titan's diversity of technology and diversity of market, in addition to its prospects for applying technology to the commercial market. "Putting my small technology prospects into the larger Titan company gave us a bigger stage on which to play and more clout in the marketplace," Baird said. "And in turn, I could see that Delfin was going to bring a lot to Titan because the two companies shared a lot of similarities, yet Delfin's customers and programs were nicely additive to Titan."[6]

Integrated Pieces

All of the 1998 acquisitions were part of Ray and DeMarco's plan to create a $1 billion government information technology business.

"Titan's acquisition activity is part of the rising wave of consolidation among defense information technology companies, particularly in the medium to large range," wrote a reporter for *Washington Technology*.[7] In the same article, Ray said that Titan was modeling itself after Safeguard Scientifics, of Wayne, Pennsylvania, and Thermo Electron Corporation, of Waltham, Massachusetts, both of which had successfully built stables of companies they had spun out while maintaining a controlling interest.

Consolidation among larger defense companies was "largely completed," Ray said. "It has

Delfin Systems, which Titan acquired in 1998, specialized in engineering, systems integration, and software for the U.S. Navy.

CHAPTER EIGHT: MAXIMIZING VALUE

gone much slower for the smaller defense information technology companies, but it is finally here."[8]

Despite the speed with which Titan completed its 1998 acquisitions, the company did not operate in a hasty fashion. Titan studied more than 100 acquisition candidates over the year, and those that made the cut had to meet stringent requirements. Ray and his team insisted that the companies they took on be a strategic and technological fit with Titan, have a solid management team, add to earnings, and have an internal culture similar to Titan's.

"There's an incredible synergy of these acquisitions that didn't come by accident," said Diane Scott, who joined Titan in 1995 as vice president of human resources. "Gene and Eric work very hard to find that synergy, and that's one reason why we've been very successful in integrating these organizations into Titan. That, and we allow these companies to operate like we're a small company."[9]

Gene Ray had always realized the value of good employees, and he made a point of showing his appreciation to Titan's workers. Part of that appreciation involved fostering a free-thinking corporate culture that inspired employees to want to do their best.

Ken Kreyenhagen remembered feeling apprehensive about how his job might change after Titan had acquired his company, California Research and Technology, in 1986. "I wasn't sure what Titan's management style would be," he said. "But looking back, I have no regrets. My relationship with Titan, with Gene Ray and the senior management team, were always very good. They were very anxious that the companies they acquired remain largely autonomous, and there wasn't a push to enforce conformity. Ray strongly felt that he was buying groups of people, and

Titan continued to support the Joint Surveillance Target Attack Radar System (Joint Stars)—a cutting-edge C⁴ISR system—for the U.S. military.

I think he wanted them to retain their independence to some degree."¹⁰

That independence, that entrepreneurial spirit, was and remains a major ingredient in Titan's corporate culture and has enhanced the company's innovative strength over the years.

All told, revenue of the five 1998 acquisitions totaled about $160 million, which more than doubled Titan's 1997 sales.¹¹ The benefits encouraged another buying streak in 1999.

More Acquisitions

Transnational Partners (TNP), a San Diego–based software company that provided infrastructure and enterprise resource planning for corporations, joined Titan in January 1999. TNP's solid business, good profits, critical mass, and proven management further enhanced Titan Software Systems, making the software systems division a likely spin-out or spin-off candidate in the future.¹²

Perhaps even more important, the TNP purchase brought David Porreca back into the Titan lineup. Porreca, who had been involved in the start-up of four companies since leaving Titan in 1988, was founder and CEO of TNP. Ray named him president and CEO of Titan Software Systems.

Porreca remembered the exhilaration of rejoining the Titan team at that point in the company's history. "Gene was going to transform Titan," Porreca said. "He had started down that path in 1996, and now there was going to be a new sense of excitement as we tried to do in the commercial world what we had done in the defense world: take on and solve tough problems of an international scope and apply our technology to more peaceful pursuits and those focused on the global environment."¹³

Titan's second acquisition in 1999 was System Resources Corporation (SRC), a closely held, Boston-based provider of information technology solutions and services. SRC specialized in aviation and airport systems, security, logistics automation, and command and control.

Atlantic Aerospace Electronics Corporation (AAEC), of Washington, D.C., was purchased in July 1999. The privately held company had been founded in 1985 by Robert Cooper, the former director of NASA's Goddard Space Flight Center, and focused on applied research and development in information technologies.

"AAEC brings to Titan a brilliant research staff and an entire library of innovative and leading-edge technologies," Ray announced. "The adaptation of its technological developments into marketable products and systems will favorably impact the continued profitable internal growth of our company."[14]

With eight acquisitions completed, Ray and his team once more restructured the company's operations during 1999, producing four main Titan businesses (and renewing some bright names from its past): Titan Systems, Titan Wireless, Cayenta, and Titan Scan—plus an Emerging Technologies and Businesses segment.

Titan Systems

Titan Systems was the corporation's stronghold. Its profitability boosted Titan's growth, or as Bud Baird liked to point out, "Titan Systems has been an engine of *profitable* growth for the company—because growth without profit or profit without growth will get you into some hot water." Titan Systems not only generated cash that fueled the growth of Titan's commercial businesses but was also the source of many new technologies.[15]

Titan's defense and government IT markets remained strong, and for the first time in more than a decade, the national defense budget was on a modest upswing.

The technologies under Titan Systems' wing included the following:

- The UHF Tunable Patch Antenna, a state-of-the-art satellite communications antenna
- Advanced wave forms, which allowed existing UHF satellite communications a significant increase in data throughput
- Interactive software tools to fight the threat of terrorism worldwide

- Real-time video and image-processing technology that could be customized for security and surveillance needs
- A Rapid Retargeting Technology that could help military customers breathe new life into aging systems and components
- "Prophet," a multistation signal-intercept and direction-finding system to support the U.S. Army in the field

In July 1999, Titan Systems increased its share of government work with a $38 million award from the FAA. Under the contract, Titan would provide support in airspace and airports analysis, simulation, modeling and development, human factors analysis, operational concepts analysis, and information technology.

Titan Wireless

Titan Wireless was essentially a commercial communications outgrowth of Linkabit, a transfer of Linkabit's expertise in satellite-based telecommunications technology—especially DAMA technology—to the commercial marketplace. It was that expertise that allowed Titan Wireless to emerge as a rapidly growing, low-cost service provider in developing countries.

"At Titan Wireless, we've developed a global network to serve emerging telecommunications markets around the world," said Mike Kulinski, who had been with Linkabit since 1984. "We use a variety of technologies such as satellite communications, fiber optics, and different wireless transmission techniques, and we provide basic communications services—which can include voice, data, and Internet services—to very diverse markets."[16]

Key elements of Titan Wireless' infrastructure included the following:

- Common network architecture and commercial, off-the-shelf components for ease of operation, maintenance, and expansion
- Centralized global network control and centralized troubleshooting, diagnostics, configuration management, security, billing, accounting, clearinghouse, and other essential support functions
- Low-cost hardware and proprietary software

Titan Wireless developed a network that served emerging telecommunications markets around the world. By the end of 1999, it delivered long-distance minutes between the United States and 10 countries in Africa, Asia, Latin America, and the Middle East.

By 1999, Titan Wireless had installed satellite-based telecommunications gateways in 10 countries in Africa, Asia, Latin America, and the Middle East.

In Guatemala, for example, Titan Wireless provided telephone service to small villages. "This gives a village that perhaps had no communications connectivity at all the ability to call anywhere in the world," said Kulinski. "There are many Guatemalan immigrants here in the United States who like to keep contact with their families back in rural Guatemala, and we provide that at a very low cost."[17]

Cayenta

Titan's commercial software integration business, renamed Cayenta, evolved during 1999 into a full-service e-commerce subsidiary that offered a complete suite of Internet-based management systems. The name Cayenta was derived from a Native American term used to describe natural rock formations in the Southwest. The name symbolized Cayenta's business: building the rock-solid infrastructures required for complete e-business solutions. In other words, Cayenta was a "Total Service Provider."

David Porreca, who had rejoined Titan that year when it purchased TNP, was the driving force behind Cayenta's growth and became Cayenta's president and CEO. "Gene told me he was going to transform Titan by creating, building, and spinning out companies and technologies," said Porreca. "That's what induced me to rejoin him, to take one of my partnerships and have it merge with Titan Software Systems to create a new company."[18]

Cayenta's vision, Porreca explained, was to bring together everything needed for a system "because multiple applications sitting on a server someplace that are fundamentally isolated islands of automation don't do anything for coherence or collaboration." For most commercial applications, human beings are the glue that seamlessly ties together the silos of information. "But in the Internet world," Porreca said, "you can't afford to have human beings tie it together; they're not fast enough. You need knowledge at your fingertips. So you have to have a membrane, a platform, of multiple applications working together as a system. And that's what Cayenta is providing."[19]

Cayenta focused on what Porreca called "four vertical areas":

- Multichannel commerce (including catalogs, direct merchandising, and retail point of sales)
- Complex utility management systems (electric, telecommunications, gas, water, etc.)
- Manufacturing logistics (supply chain logistics and inventory)
- National enterprise systems

The acquisitions and joint ventures that Cayenta made enhanced one or more of these vertical areas. In September 1999, for example, Cayenta and the publicly traded Sempra Energy formed a joint venture called Soliance Network to provide Internet-based information services to help midsize utility companies adapt to deregulation.

In November and December 1999, Titan acquired two other companies that broadened Cayenta's offerings. The first, J. B. Systems (also known as Mainsaver), provided enterprise asset management software and distributed workflow management technology. The second, Solutions for Growth, provided revenue cycle management software for companies with complex bill calculation requirements.

Cayenta's mission was to change the way that applications are delivered to companies. But Cayenta also sought to become a successful public company on its own, which was, of course, part of The Titan Corporation's overall vision. Cayenta came close to realizing that goal on December 29, 1999, when it registered with the Securities and Exchange Commission for an IPO. The plan was to sell a minority interest to the public while majority ownership would remain with Titan. In addition, Titan would execute a tax-free spin-out of its remaining interest directly to Titan shareholders. This would mark the first public offering of a Titan subsidiary.

Unfortunately, the timing for the spin-out wasn't right. The dot-com bubble had burst just before the scheduled roadshow, and stocks in general, especially technology stocks, were taking a nosedive. "We don't seem to have good luck in terms of getting our spin-outs ready," said Virginia

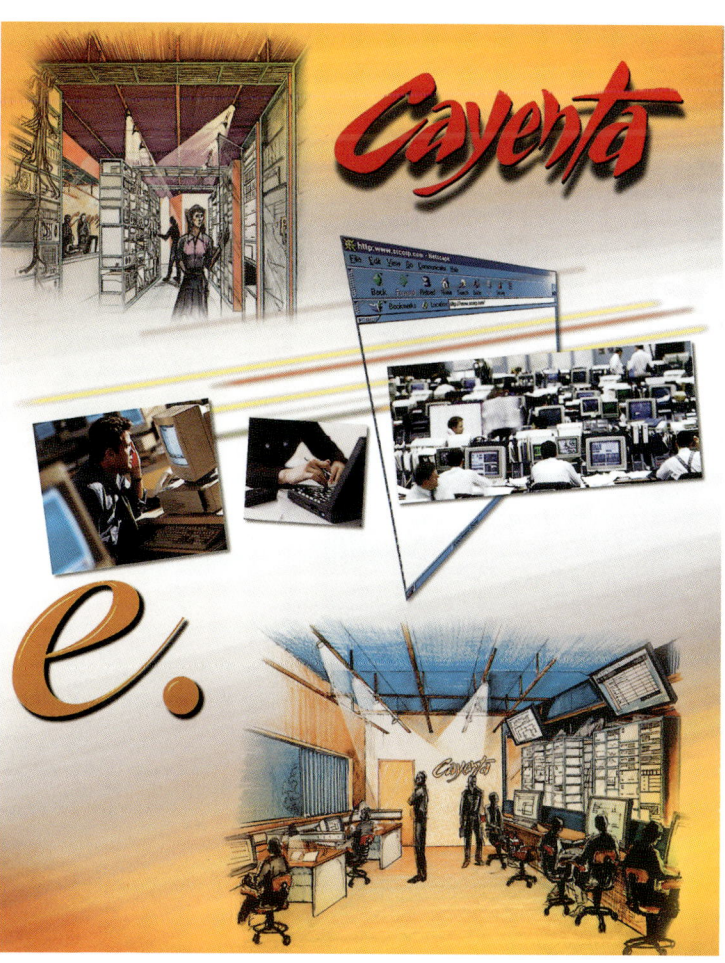

Cayenta, Titan's commercial software integration business, was a full-service, business-to-business e-commerce company that earned a reputation for being a "Total Service Provider."

Oliver, assistant corporate controller. "Just when we're ready, it seems the market has a bad turn."[20]

But the scrubbed IPO didn't deter Cayenta from its mission, which Porreca called the "great entrepreneurial opportunity to invent the future."[21]

Titan Scan

During a year punctuated by many noteworthy business achievements, Titan Scan's revolutionary SureBeam® technology reached new heights. Both the Denver and the San Diego medical sterilization facilities were upgraded during the year, but it was SureBeam's application to food that captured headlines and worldwide interest as none of Titan's other products had.

According to data from the Centers for Disease Control and Prevention, food-borne bacteria, often found in meat, killed more than 5,000 Americans a year and sickened up to 76 million. Meat distributors in the United States were in a heightened state of alert regarding *E. coli* 0157:H7 bacteria and were searching for a way to eradicate the pathogen.

The FDA had approved the use of electron beam irradiation to kill harmful pathogens in meat in December 1997. But it wasn't until early 1999 that the USDA released proposed rules for the use of irradiation to sterilize meat, and it would take the agency until December 1999 to actually approve the process.

SureBeam®—designed to kill *E. coli*, listeria, salmonella, and other food-borne pathogens—was way ahead of the game. In April 1999, Titan began construction of the first SureBeam system capable of pasteurizing ground beef at a facility in Sioux City, Iowa. The system would be capable of processing more than 250 million pounds of meat product a year at a cost of about one penny per hamburger.

By the end of 1999, companies that produced 75 percent of the country's ground beef and approximately 50 percent of the poultry had agreed to test SureBeam technology at Titan's Sioux City facility. These companies included Cargill, IBP, Tyson, Emmpak, and Huisken Meats.[22]

"Titan's the leader in electron-beam technology," Mark Klein, a spokesman for Cargill, told the *San Diego Business Journal* in April 1999.

"We really feel obligated to use the best technology for enhancing the safety of our products. You really can't put a price on food safety."[23]

Moreover, the SureBeam technology, because it carefully controlled the electron beam, did not diminish the taste or quality of the food. Tom Allen explained how Titan had perfected the science before patenting the technology.

The science of killing a specific organism is well known, and there's not much you need to do to add to that science, but when you place that organism in a moisture environment of nitrogen or high oxygen, all of these external factors influence the dose at which you can effectively kill the organism and not change the sensory characteristics of the product. The higher dose I give it, the easier it is to kill the organism, but I also run the risk of changing the quality characteristics of the product. Today we have a much better understanding of how to apply the technology for the proper depth dose distributions.[24]

Finally, in December 1999, the green light that Titan had been waiting for blinked on when the USDA approved the use of electron-beam technology to kill bacteria in meat. The news sent SureBeam's new food processing facility in Sioux City into full throttle, and Titan's stock leaped more than 12 percent.[25]

In the glow of SureBeam's success, Ray noted that his managers had suggested building a food

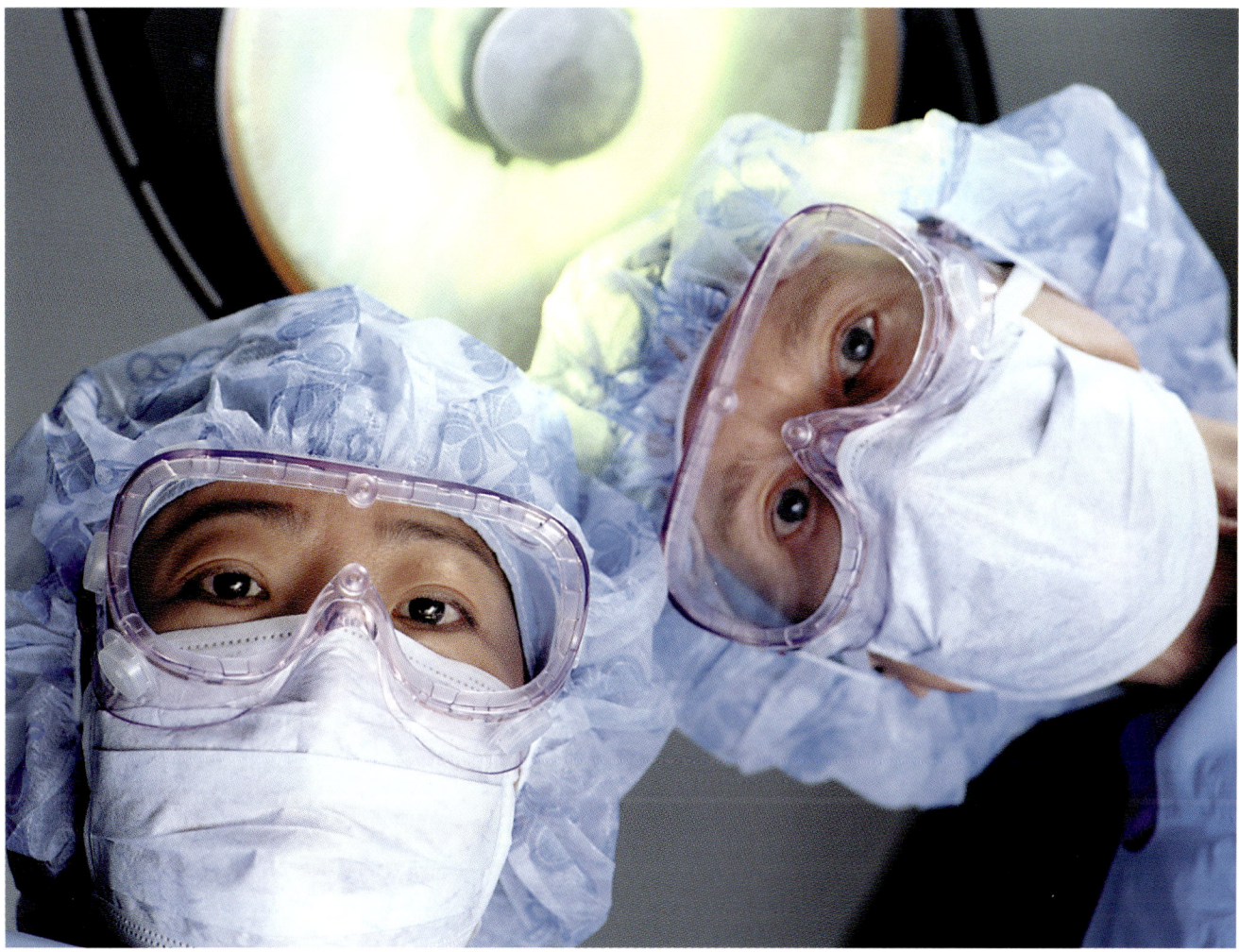

pasteurization plant 16 years before, when the company's electron-beam technology was being used for Star Wars. "I thought we might be a little ahead of our time," Ray said. "Now the time has come. All along we knew the market was going to mature."[26]

The market did start to materialize, mainly due to continued problems with food safety. Hawaii had been unable to ship fruit to the continental United States without quarantining because of a statewide problem with fruit flies. Methyl bromide, a pesticide used in many parts of the world to keep fruit flies and their larvae from spreading from one country to another, was found to be harmful to the earth's ozone layer. So the state's agriculture industry had been destroying the pests by treating the fruit with hot water vapor, but that method left papayas dehydrated and unripe. A couple of entrepreneurs had suggested building a cobalt-60 plant to irradiate the fruit, but that idea met with tremendous local opposition.[27]

"Then they noticed that we were moving forward with our SureBeam® technology," said Larry Oberkfell, a Titan senior vice president and president and CEO of SureBeam Corporation. "They said, 'Hey, here's a way for us to irradiate the papaya with electricity, kind of like a microwave oven without the heat, and not have to worry about any of the nuclear issues. And overnight the opposition was gone.'"[28]

Titan Scan uses SureBeam's patented electron beam technology to sterilize medical devices and equipment. Because the technology uses ordinary electricity for sterilization, it is more efficient than other sterilization methods and poses no environmental risks.

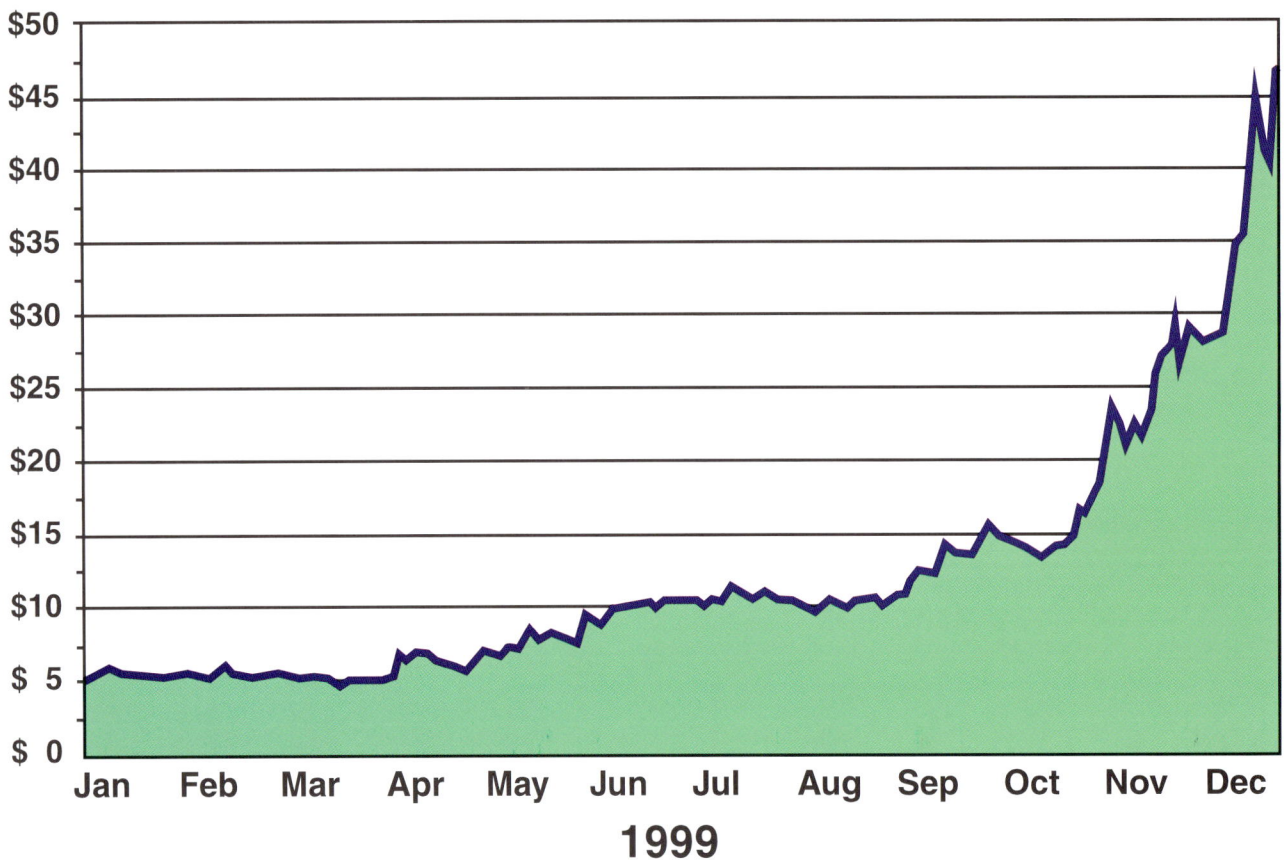

A Year of Steady Growth
1999

In the latter part of 1999, Titan's stock gained dramatically in value. In January it hovered at around $6 a share, but by the end of the year it had surpassed $46, making it the largest gainer on the NYSE. That growth represented a gain of 760.2 percent—more than 250 percent greater than the second-place winner.

In late 1999, Titan Scan formed a joint venture with Hawaii Pride to install a unit near Hilo, Hawaii, that would use SureBeam's x-ray system to disinfect flowers and papaya and other exotic fruits.[29] (The x-ray system based on electricity was essentially the same as the electron-beam system but allowed greater penetration to treat thicker products.)

The joint venture would ship the first SureBeam®-processed fruit to the mainland in August 2000, and by August of the following year the partnership was shipping 750,000 pounds of papayas a month from Hawaii to the continental United States. Titan's 19.9 percent interest in the joint venture allowed for ongoing profits, and as Oberkfell observed, "It really rejuvenated family farming on Hawaii."[30]

SureBeam could help poor, agrarian-based countries improve their economies and quality of life by making it possible to export more food products to the global market. "The banning of the methyl bromide pesticide had left much of the world, in terms of international trade, devoid of any kind of treatment, and those countries therefore faced a potential loss of international trade," explained Dr. Denny Olson. "So looking at what alternatives are available, it's clear that food irradiation is the best way to control insects and pests in fruits and vegetables for international trade."[31]

In addition to destroying harmful pathogens and insects on food, SureBeam could actually extend the freshness of food from eight or 10 days to 20 days. This benefit had a dramatic economic impact on farmers, who would no longer have to ship fruit and vegetables by air. Now they could save immense amounts of money by shipping the product less hurriedly by sea.

Allen pointed out that SureBeam also allowed countries with underdeveloped transportation systems to more effectively distribute food within their borders. "We're completely changing the way companies look at the economics of distributing food," he said. "We had a session with a Chinese delegation that had indicated they were losing 40 percent of their grain by the time they got it distributed from the ports to where they actually consumed the grain. Well, irradiating the product can kill a lot of the spoilage organisms that actually destroy the grain."[32]

Restaurants, too, could benefit from the longer shelf life of food. "If a fast-food chicken company is getting three deliveries of fresh chicken a week, they can now get two," Oberkfell explained. "They haven't just eliminated one delivery. They've eliminated a third of their whole distribution cost."[33]

Titan Scan formed another joint venture company in 1999, this time with Zero Mountain, one of the nation's largest independent suppliers of cold storage services for poultry. The joint venture purchased a SureBeam unit to offer irradiation services to users of a Zero Mountain cold storage facility in Russellville, Arkansas.[34]

A Powerhouse

All in all, 1999 was a magnificent year for The Titan Corporation. Its common stock was the number one market gainer on the New York Stock Exchange, and its market capitalization increased to $2 billion.

Titan continued to grow bottom-line earnings per share by more than 30 percent while making significant investments in its rapidly growing commercial businesses. Titan Wireless had seen 300 percent revenue growth, and Cayenta's revenue increased by 100 percent.

Titan's mission to "create, build, and launch technology-based businesses and then maximize Titan shareholder value by spinning these businesses off as stand-alone public companies" was truly being realized.

Initial Public Offering of SureBeam®
New York Times Square
March 16, 2001

SureBeam was the first Titan company to spin out as an independent company. On March 16, 2001, it became a publicly traded company on Nasdaq.

CHAPTER NINE

TITAN TO THE RESCUE

2000–2002

If you hold shares of Titan Corporation, you are probably one happy investor right now. The stock was the biggest percentage gainer on the New York Stock Exchange last year.

—Reporter Lauren Thierry,
CNNfn, January 4, 2000

CNNFN'S BUSINESS BROADCAST ON January 4, 2000, featured the chairman and CEO of "a little-known company," but one "we'll know a lot more about in the year to come," broadcaster John Metaxas told viewers.

The honored guest was none other than Titan's Gene Ray, who was asked to share with the CNN audience the secrets behind his company's phenomenal stock performance in 1999. Though Titan might not have had a famous name, it had achieved some famous numbers, with an extraordinary 760 percent gain in its stock price during the year.

"Titan can be described as a technology incubator that incubates technology-based businesses," Ray told viewers. "Our strategy then is to grow those businesses to the appropriate level, spin out a minority interest to the public, . . . and then, at the appropriate time, to spin those businesses off tax-free to our shareholders."

The business blueprint Ray described was one that would not only bring more fame to Titan over the next couple of years but successfully drive the company to the accomplishment of Ray's long-term goal—the billion-dollar revenue mark. The company's stock had skyrocketed in 1999, and in 2000 Titan would become a billion-dollar corporation with revenue up 154 percent from the previous year.

This amazing success would not be achieved without some highs and lows. While Titan executives would be delighted to find themselves in the national press limelight during some key historic moments, the company would also struggle through some unflattering rumors and a volatile stock market. As always, The Titan Corporation emerged from such challenges stronger and more determined than ever.

Joining Pieces to Titan Systems

As Titan entered the new millennium, Titan Systems Corporation, the company's government information technology (IT) subsidiary, remained the corporate cash engine and supplied most of Titan's new technology. The division's 2000 revenue was $796 million, more than double the previous year's $311 million.

Titan Systems was able to reel in ever larger contracts thanks to its expertise and Titan's aggressive acquisition strategy. It was as though Titan's

---

Titan acquired Datron Systems in 2001. Among Datron's many products was the Guardian conventional portable radio. Capable of use in government, military, public safety, and business operations, the radio is fully digital and meets APCO Project 25 standards for compatibility with other Project 25 radios.

leaders were rounding out a masterwork, searching for companies that were a strategic fit, buying them, and joining them with Titan Systems to enhance technology and systems and increase profits.

One of Titan's biggest buys was announced in January 2000. Titan was to pay $175 million in stock for Advanced Communications Systems (ACS), of Fairfax, Virginia. ACS had a proven track record of providing communications, intelligence, surveillance, and reconnaissance services to the government and corporations. This acquisition brought 2,000 new employees and opened the door to another $218 million in annual revenue. What's more, because the deal was consummated with a stock trade, it left Titan unfettered with debt and free to make further acquisitions.

Analysts heartily approved the move. "Superlative," said Mark Jordan, an analyst with the investment banking firm A. G. Edwards, of St. Louis. "ACS is one of the best companies within that marketplace."[1]

Others agreed. "Gene Ray has done a wonderful job of making all the elements of the business work," said Richard Knop, a partner in the investment banking firm Boles, Knop & Company, of Middleburg, Virginia. "The public markets have been rewarding him, and they are clearly on a growth track."[2]

"They really are building a high-end, defense IT and communications business," said Jerry Grossman, a director at the investment banking firm Houlihan Lokey Howard & Zukin, of McLean, Virginia. "They have really been aggressive and determined."[3]

Four more acquisitions followed on the heels of the ACS purchase.

LinCom Corporation, a Los Angeles–based developer of wireless communications and information systems, was acquired for cash in February 2000. Founded by Dr. William C. Lindsey, a renowned pioneer in digital wireless communications, LinCom had been developing communications and software technology for NASA space programs and U.S. military satellite programs for 28 years.

Ray pointed out the synergies between Titan and LinCom. "We have a long history of providing cutting-edge satellite-based wireless telecommunications systems," he said. And because LinCom focused on the next generation of commercial wireless mobile communications, it added "a valuable new dimension in wireless communications" to Titan's technology base.[4]

Pulse Engineering, a privately held, Maryland-based IT company, became part of Titan Systems' stable in March 2000. Pulse, which had been founded in 1979, brought to Titan expertise and technology in the areas of strategic and tactical information security services and products as well as engineering design, development, integration, and test and deployment of signal processing systems. The purchase also brought 180 new employees, which strengthened Titan's relationship with and role in the intelligence community.[5]

In the summer of 2000, Titan announced the acquisition of SenCom, of Bedford, Massachusetts, a provider of IT services to the U.S. Air Force, the Defense Information Systems Agency, and the Customs Service.

AverStar, a Burlington, Massachusetts, IT company that focused on information assurance, infor-

Gene Ray, cofounder, chairman, and CEO, has led The Titan Corporation to unprecedented success.

In Memory

MELLON C. "BUD" BAIRD, CEO AND president of Titan Systems, passed away on March 5, 2002. He was 71. Despite his battle with myelodysplasia, a blood disorder, Baird remained actively involved in Titan until one week before his death. For many who knew him, his death was unexpected and, of course, a devastating blow to his family and friends and to Titan. His more than 40 years of experience in the defense and aircraft industries contributed greatly to the defense and security of the nation.

For his funeral, his family requested that, in lieu of flowers, people make contributions to the M. C. Baird Fund at the University of Texas's College of Engineering.

mation operations, and network and information security, was yet another Titan acquisition for 2000. AverStar's customer base included the Department of Defense, civilian government agencies, and commercial businesses. Joseph Saponaro, president of AverStar, joined Titan Systems as president of the Civil Sector. Michael Alexander, AverStar's chairman and CEO, joined Titan's board of directors.

With annual revenue of approximately $230 million, AverStar was the buy that pushed Titan past the $1 billion mark—and kept analysts talking.

"Titan Corp. has vaulted itself into the upper tier of government information technology providers," reported *Washington Technology*, noting that Titan's acquisitions had positioned it to compete "head to head" with such illustrious companies as Affiliated Computer Services, Computer Sciences, Litton-PRC, Lockheed Martin, Logicon, and Ray's one-time employer Science Applications International. Or, as another analyst pointed out, "They have leapfrogged over the other middle-tier companies."[6]

One analyst noted that Titan would have some work to do to meld the various businesses.[7] Titan had become a master, however, at integrating new businesses into its infrastructure and corporate culture.

Furthermore, because of the strong synergy among Titan's divisions, the company was able to win important contracts that involved collaboration of products and services from multiple Titan Systems divisions. As Bud Baird, president and CEO of Titan Systems, pointed out, "A strategic fit has always been a company criterion for acquisition, but now it is even more important because we have a shape to the market areas that we wish to pursue."[8]

Charles "Chuck" Saffell, who joined Titan Systems from SAIC in late 1998 as vice president for C^4ISR (command, control, communications, computers, intelligence, surveillance, reconnaissance), worked with Baird to increase and enhance Titan Systems' multiunit opportunities. "Bud Baird had the vision," he said. "If you're going to acquire these new units, you want to acquire a synergy by combining the capabilities of new units with each other and with existing capabilities within the corporation. That way you can compete in a market that requires greater bulk in size and capability and expertise."[9]

All told, Titan's year-2000 buying spree left it with a total of more than 1,000 active contracts and a contract backlog in excess of $2 billion.[10]

Conquering Rumors

Beginning in the summer of 2000, Titan found itself having to conquer some ugly rumors. As far as anyone could tell, the problem began on the Internet. Using a Yahoo bulletin board, anonymous messagers began spreading rumors that Titan was facing major problems and that investors were bailing out.

"Oh My Gosh!!!!!!! TTN Titan is getting nailed with huge sell orders!!!! Jump from the sinking ship!!!!" wrote one messager.

"Institutions are selling big time!!!!!" wrote another.[11]

The messages were not factually correct and included bogus reports of poor company earnings and fraudulent accounting by Titan. In addition, a San Francisco–based asset-management hedge fund, which made its living selling company stocks short, circulated a report on Wall Street that suggested Titan was inflating its profits through questionable accounting practices.

Though none of this was true, Titan watched the price of its stock drop from $44 in May to $21 in August. Titan executives and analysts blamed the rumors on the hedge fund. "There is no legitimate reason for the drubbing the stock has gotten," said analyst Mark Jordan, who had been following Titan for three years.[12]

Titan filed a lawsuit against the hedge fund, accusing it of driving down Titan's stock price in order to make a profit of short selling. In a settlement agreement, the firm later acknowledged that Titan's accounting practices were not faulty. The terms of the agreement were not disclosed; however, Titan seemed very pleased to drop its lawsuit.

The stock rebounded by November 2000, after investors had a look at Titan's third-quarter results. One local investor told the *San Diego Union-Tribune* that he was "very glad that Titan has been vindicated. It's hard enough for a company to face the current conditions in the stock market without having somebody purposely spreading inaccurate bad news about them."[13]

A Better (and Bigger) Shape

Titan continued shaping its market segments in 2001 with more strategic acquisitions. The publicly traded Datron Systems Incorporated joined the Titan fold in August. Founded in 1969, Datron served both commercial and military markets. It made satellite antennas to track airborne weapons and produced voice- and data-communication radio products. All of its radio products were known around the world for their ease of operation, reliability, serviceability, and affordability. Datron was headquartered in Vista, California, and had plants in San Diego and Simi Valley. Its 300 employees became part of Titan Systems Corporation.[14]

Then in November, Titan bought BTG, Inc., based in Fairfax, Virginia, just outside Washington, D.C. BTG provided systems and solutions development, analysis and consulting, and integration and support services to U.S. defense, intelligence, and government customers. The company specialized in information collection and analysis, warfare modeling and simulation, software and systems integration, network design and architecture, and information and network security. Its 2,000 employees also joined Titan Systems.

BTG was founded in 1982 by Dr. Ed Bersoff, a mathematician who wanted to service the military intelligence market. The company's innovative rapid prototyping technique, which it developed for its first contract, quickly gained attention in the military intelligence community. The company expanded that work into other areas and started building practices across the entire military and eventually the whole government spectrum.[15]

BTG grew rapidly. It went public in 1994 and executed a follow-on offering in 1997. In 1999, BTG entertained the notion of being acquired and put the word out. "But we ended up not doing it for a variety of reasons," said Bersoff. Then in 2001, Titan approached the company and asked if it would be interested in dealing directly with Titan. "We started that conversation and ended up with a transaction about four or five months later," said Bersoff.[16]

Bersoff noted that Titan's and BTG's markets "kind of touch near one another, but they largely don't overlap. In very few cases are we in pursuit of the same market, so the deal is very attractive for both of us." And, of course, the joining of Titan and BTG enabled both companies to bid on larger programs.[17] Indeed, an analyst with Merrill Lynch predicted that the synergies between the companies should add 5 to 6 cents a share to Titan's earnings within the first year.[18]

Titan Systems

Bud Baird described Titan Systems as "providing innovative solutions to complex problems."[19]

Specializing in C^4ISR systems, it drew about two-thirds of its $919 billion 2001 revenue from defense and intelligence work and the other third from civil government agencies such as NASA, the FAA, the Department of Justice, and the Drug Enforcement Agency. Its unique capabilities in sophisticated data management, information processing, information fusion, and knowledge-based systems and communications made it a leader in providing information systems solutions and services to the U.S. government.

Baird saw a strong future for Titan Systems. Even before the terrorist attacks of September 11, 2001, he noted that "World events have created a need for more and more sophisticated responses in order to protect the country, and that presents wonderful opportunities for our company." Just as the end of the Cold War had presented Titan Systems with opportunities for commercial technology and technology that helped stabilize potential conflicts all over the globe, Baird knew that Titan would "come up with new ideas, new concepts, new architecture for dealing with the world as it may be 10 and 20 years from now." And, he pointed out, "Titan Systems is in the business of helping our customers work on such architectures."[20]

"We're a notch above just a plain old defense company," Saffell said. "We help the government develop technology, then take it to the commercial sector and let the commercial sector make it more robust. And then, because we uniquely understand the technology, we can bring it back to our government customers as commercial off-the-shelf technology and save them a great deal of money and time—because now they don't have to go through all the very detailed development cycles that the government is required to go through if it makes something unique."[21]

Though each of the divisions that made up Titan Systems had unique specialties, they often

Titan Systems' C^4ISR systems can be found in such aircraft as the P-3 Orion, the Navy's long-range surveillance aircraft.

Titan Systems provides unique capabilities in sophisticated data management, information processing, information fusion, and knowledge-based systems and communications requirements.

worked together on innovative projects. "We have one database where we keep all of our potential opportunities," said Saffell. "Each group has access to that database. So if I were to see a potential opportunity, I could search the database and see if anybody else was already tracking it, and if they were, we could decide who was best able or capable to lead it. And then we'd work together as a multisector, multigroup team and go after that opportunity."[22]

Titan's broad span of expertise, in turn, allowed it to offer its customers a more complete package and to compete for larger contracts.

Engineering and Information Services

At the highest level, Titan Systems could be broken down into two major segments: Engineering and Information Services, which was a service-based business, and Science, Technology and Products, which focused on bringing products and technologies to customers.

"The value proposition we take to our customer is that we can solve your problem," said Saffell of the Engineering and Information Services segment. "We come up with a solution. We're able to integrate that solution into the customer's legacy system or their existing capabilities. We provide the people—everyone from welders who can cut a hole in a ship to those who can install a system in an aircraft. We also have people who work the integrated logistics of the system, those who provide training in the new system, and those who develop the supply and support apparatus to maintain it."[23]

The Engineering and Information Services segment brought in more than 60 percent of Titan Systems' revenue and comprised three sectors—Civil, Technical Resources, and Maritime—and each sector was broken into groups and divisions.

The Civil Sector

The Civil Sector, led by Joe Saponaro, who joined Titan in 2000, when it acquired AverStar, employed about 2,100 as of May 2002 and brought in about $270 million in yearly revenue. The Civil Sector comprised two groups: Civil Government Services, with Bruce Burton as group president, and Applied Engineering Solutions, led by group president Les Rose.

The Civil Government Services Group provided IT services and e-government solutions, mainly to non-DoD and nonintelligence federal government agencies. Its primary customers included the Department of Health and Human Services, NASA, the U.S. Postal Service, the Patent and Trademark Office, the Federal Deposit Insurance Corporation, the Federal Communications Commission, the Bureau of Labor Statistics, and the Environmental Protection Agency. It also provided IT solutions to state and local governments such as the Commonwealth of Massachusetts and the state of Illinois. In 2002, the group employed nearly 1,000 professionals, who provided project management, IT assurance, system development, and IT outsourcing from more than 20 locations on three continents.

The Civil Government Services Group's Civilian Space Division continued to provide systems assurance, systems engineering and development, and project management. For example, it independently verified and validated all the major NASA systems at Cape Kennedy—for the space station, the space shuttle, robotics, and launch processing. Another division, Astronautics Engineering Operations, built some of the real-time cockpit simulations for NASA Johnson Space Center.

The Health and Social Resources Division offered systems assurance, program management support, systems engineering development, and geospatial services to government agencies devoted to health, education, and the welfare of citizens. It provided systems assurance services to the Centers for Medicare and Medicaid Services, for example, to help ensure that millions of Americans received their Medicare checks on time.

The IT Outsourcing Division specialized in IT infrastructure management and IT outsourcing services. The Patent and Trademark Office was a major customer of IT Outsourcing. About 200 people from the division supported 6,000 employees in that office, running the networks and help desks and providing around-the-clock service.[24]

IT Services & Solutions, another division, delivered e-government solutions to help agencies take advantage of new technologies while assuring the

Titan Systems is a leading provider of information technology services and has earned numerous awards for its contributions to assuring mission success of some of the country's most critical and complex information systems.

integrity and security of mission-critical information such as payroll and personnel data and intelligence and military plans. It could help customers set up security systems or monitor their networks for intrusions, for example. This service became especially important after the terrorist attacks of September 11, 2001. As Saponaro pointed out, "More and more the government will need e-government solutions, both for homeland defense and for secure networks. That represents good opportunities for us to grow and compete in the marketplace."[25]

Finally, the Financial and Regulatory Division provided systems assurance, testing, systems engineering, development and integration, software maintenance, and project management support to government financial and regulatory agencies. For example, the division continued to test and evaluate all the U.S. Postal Service's automation systems, helping to ensure that citizens' mailing costs are accurate.

The Applied Engineering Solutions Group, also part of the Civil Sector, offered network management, database development and data migration, Web application development, help desk operations, and other services. It provided cost engineering and cost management as well as technology-based training. For example, it offered a course called Operations Security Fundamentals, which taught corporate and intelligence communities the importance of operations security and how to identify potential weaknesses in that security. It also helped K–12 school districts bring technology into classrooms by providing everything from initial needs assessment to technology audits, design, management of day-to-day operations, and help desk support. College-bound students could register for aptitude tests and scholarships using the division's e-commerce solutions.[26]

The Applied Engineering Solutions Group offered a wide variety of general consulting services, including a unique service that used a computerized method of developing construction budgets for federal, state, local, engineering, and architectural organizations. This was done through predeveloped engineering methods that estimated building costs and actually built a building within the computer program. Through cost estimates in the early phases of project development, the system helped clients stay within an established budget. It allowed design errors and omissions to be avoided and could quickly analyze possible alternatives and cost impacts of design choices. If, for example, the state of Florida wanted to change the facade of a new school from stucco to brick, the cost impact would be instantly apparent.[27]

The Applied Engineering Solutions Group also provided graphical information system solutions to a variety of federal and commercial customers through raster- and vector-based data and digital imagery. To address environmental concerns, for example, it offered environmental planning, natural resources management and planning, environmental compliance, and cultural resources. Its spatial models also helped solve problems and identify the most cost-effective method for managing physical infrastructures, facilities and runways, environmental cleanup, and regional land use planning. Its highly sophisticated sensors could locate, delineate, and quantify rangeland, environmentally protected plant species, and areas of licit/illicit crop cultivation.[28]

Titan Systems offers information technology solutions to K–12 school districts as well as government agencies.

The Technical Resources Sector

Earl Pontius, who had been president and CEO of Horizons Technology when Titan acquired it in 1998, led the Technical Resources Sector, which had yearly revenue of more than $400 million in 2002 and employed about 3,000 from more than 100 offices and customer sites in the United States and overseas. The Technical Resources Sector comprised three groups—Information Solutions, Defense & Intelligence Systems, and Operations, Analysis & Training—that shared a purpose of providing management consulting, systems engineering and integration, and information technology solutions to the DoD and other federal agencies.[29]

The Information Solutions Group, also led by Pontius, provided information technology, systems engineering and technical support, and management services and solutions to the DoD, federal civilian agencies, and state governments. Its customers included all the branches of the U.S. military, the FAA, and the Department of Transportation.

One of the Information Solutions Group's divisions, Systems Management Services, specialized in IT, systems engineering, technical assistance, and systems acquisition life-cycle services, mainly to the DoD and the FAA. The division supported the Joint STARS program, for example, by providing flight test engineers to act as aircraft mission crew members and ground support personnel during test missions. The FAA and Air Force relied on the division's air traffic management services, which included airport/airspace planning and design and engineering for surveillance radar programs. The division also provided physical security for the DoD, the Olympic Games, and the Bosnian Peace Talks, as well as many other systems management services.

"Since the September 11 terrorist attacks, the FAA has taken on new challenges with security," said Pontius. "Our people specialize in analyzing a security problem, developing a systems architecture, and engineering solutions using off-the-shelf equipment."[30]

Another division, called Test and Evaluation, provided test and evaluation, systems engineering and integration, and life-cycle support for IT advancement. Its major customers included the U.S. Defense Information Systems Agency, the Army, the Navy, the Coast Guard, the National Imagery and Mapping Agency, and the Defense Logistics Agency.

The Air Traffic Systems Division provided IT research and development, system solutions, and technical support for air traffic control, navigation, meteorology, law enforcement, social services, and health care automation. Major customers included the Department of Transportation, the FAA, and NASA. For example, the division supported the FAA headquarters in Washington, D.C., by providing engineering and management support. It also supported the FAA's Technical Center in Atlantic City, where it conducted advanced research and testing in simulation and modeling.[31]

The Information & Logistics Support Division provided information systems development, systems engineering, technical services, C^4I applications, logistics automation, and automatic identification technology. It supported a number of U.S. Army locations—at Fort Belvoir, Virginia; Fort Monmouth, New Jersey; Fort Lee, Virginia; and Friedrichsfeld, Germany—along with many other customer locations.

The SenCom Division provided mission-critical IT and management support services to government agencies. For example, it provided technical, engineering, and management support to the System Program Office of the Airborne Warning and Control System (AWACS) and provided program management and technical and operational support to the U.S. Air Force Tactical Exploitation of National Capabilities Program Office, which supports space-based technologies for the warfighter. The division extended the Information Solutions Group's reach into the Air Force by providing radar engineering and communications engineering expertise. It provided technical staff for the Electronic Systems Center at Hanscom Air Force Base, for example, doing engineering tests, logistics, and various business acquisitions support.

The Defense & Intelligence Systems Group, also part of the Technical Resources Sector, was headed by Bob Osterloh and was made up of four divisions: National Intelligence, Information Operations, CINC Support (dealing mainly with joint operations of military service organizations), and C^4ISR. Its principal mission was to support military and federal services intelligence organizations such as the Central Intelligence Agency, the

National Security Agency, the National Reconnaissance Office, the Defense Intelligence Agency, and the National Imagery Organization.

The Air Force was a major customer for C^{4}ISR solutions, which was a main strength of the Defense & Intelligence Systems Group. "There is going to be a lot of emphasis placed on C^{4}ISR in the next few years," said Pontius. "There's not a lot of money going into new platforms; the money is going to integration—tying the systems together to be a more effective fighting force—and that's exactly where we're centered."[32]

The Army, too, was going through a transformation, said Pontius, "and a lot of it is based on information technology. We have a very, very strong effort underway to support the U.S. Army's transformation goals in providing underpinning in what they do in the information systems arena—managing their data and improving their processes to bring on their new systems."[33]

The Operations, Analysis & Training Group, headed by Kevin Wilshere, offered modeling, simulation, training, and technology support for emergency management and defense programs. It developed modeling and simulation scenarios for training military forces, for example, and offered war-gaming support for such customers as the Marine Corps and the Strike Command.

"We develop the architectures, the models, and the various evaluations and apply these to the operational training requirements of the services," explained Pontius. "It's comprehensive, and it's becoming more and more important because live testing is too expensive. Much of the testing community is now going to modeling to reduce cost."[34]

As president of the Technical Resources Sector, one of Pontius's chief goals was to bring the complementary capabilities of the different divisions and groups together in order to support existing customers and pursue new businesses and larger contracts. "Many of our companies used to be competitors, and it will take some time to change that competitive mindset," said Pontius. "But there's a

Titan Systems specializes in C^{4}ISR systems that support such programs as AWACS (Airborne Warning and Control System), which provides surveillance, command and control, and early warning detection and tracking of air targets at extended ranges. This E-3 Sentry is commonly known as "AWACS" because it is the primary Airborne Early Warning aircraft of the NATO alliance.

great spirit of cooperation here, and it's contagious. One of the first things I do when I integrate a company is to make sure people understand that we work together at Titan."[35]

The Maritime Sector

The Maritime Sector of Titan Systems Corporation, led by Chuck Saffell, provided engineering and information services to maritime and intelligence customers. The Maritime Sector comprised four groups: Aviation Engineering, Communications Engineering, Integrated Installation & Engineering, and Systems Services.

The Aviation Engineering Group, with Tom Brennan as group president, provided a wide range of aviation engineering services, solutions, and technical assistance, mainly to government customers, especially the Navy and Marine Corps. It had three divisions: Aviation Integration, In-Service Support, and Naval Systems.

In support of aerospace programs, the Aviation Engineering Group provided systems engineering and design, as well as engineering and logistics services. It also supported the acquisition, upgrade, and rework of aircraft structures and systems.

In addition, the Aviation Engineering Group provided a wide range of information systems services, including design, management, integration, software programming and applications development, enterprise information portals, and relational database development for program management. The group also provided system-level survivability solutions and engineered a wide range of weapons, sensor, surveillance, air and waterborne acoustics, and reconnaissance systems. It analyzed aircraft and ship structures, applied advanced corrosion control methodologies, and supported the development of advanced materials. It supplied a full range of aviation-related services, too, including materials, structures, avionics, acoustics, personnel, and propulsion and system engineering.

"We provide engineers and technicians to support myriad functions that are vital to naval aviation," said Saffell. For example, the Aviation Engineering Group ran and managed the Navy's master database that tracked all its aircrafts' landings, takeoffs, cycles, and carrier landings by each aircraft's bureau number.[36] The group supplied a number of people who worked in survivability and vulnerability laboratories to determine fatigue life of components. The group also helped develop simulators and training systems, analyzed weapons effects, provided aid to civilian and military mechanics, and performed a number of other engineering services.

The Communications Engineering Group, led by Dan Donoghue as group president, had two divisions: Engineering and Management Services and Systems Technology. The group provided a wide range of communications-oriented technology services and solutions to government agencies and commercial and international customers, though its primary customer was the Space and Electronic Warfare Systems Command.[37]

Its communications services included designing, developing, integrating, and installing communications solutions for landline, wireless, and satellite communications, principally for the Navy. It also provided technical, systems engineering, acquisition, programmatic, and financial management support for large-scale, unique C^3 systems.

In addition, the Communications Engineering Group supplied information systems security, integration, software programming and applications development, geographic information systems, imaging solutions, information security and assurance, and relational database planning and development. For example, it provided systems and software engineering and analysis for NATO's Maritime Command and Control Information System.

It also developed mission-critical software and engineered a wide range of C^3 systems and information systems. For its commercial customers, Communications Engineering provided end-to-end solutions, including installation. For its government customers it offered a full range of integrated C^4ISR or information system services.

The Integrated Installation & Engineering Group consisted of the Integrated Services Division and the Unidyne Division and was led by Dave Conner, group president.

"These are the people who do shipboard systems integration," said Saffell. "They do all the maintenance in the hull of the ship. They do electric repair. They do photoelectronic and information systems installation. If there's a new weapons system or command and control system that needs to be installed aboard a ship, this group would make that installation, doing everything: cutting a hole in the ship, rigging the equipment, building the foundation, doing the electrical work, and welding the hole again."[38] The group kept teams of people on standby—called flyaway maintenance support teams—that it dispatched all around the world to service the equipment that Titan specialized in.

The Integrated Installation & Engineering Group specialized in installing antennas, "and it's not just hanging an antenna on the side of the ship," explained Saffell. "It has to be coordinated with the other radiating antennas aboard the ship."[39]

In addition, the group installed training systems that were integrated into a ship's combat system, allowing the Navy to train its sailors without actually going to sea.

The Systems Services Group, with Al Branch as group president, provided technical and engineering services to the DoD and U.S. intelligence agencies such as the National Reconnaissance Office (NRO). Like the Integrated Installation & Engineering Group, the Systems Services Group installed systems, "but it's more oriented to the factual metrics, the software, that controls the systems rather than the hardware," explained Saffell. "It's still installation, but it's installation of a program load for combat systems or an integration of two systems that have to talk to each other."[40]

The Systems Services Group also developed software tools. For the NRO, for example, it developed a software tool called Intelliscape. "This is an analysis system that allows you to look into many different databases and be able to pull the information together to a logic tree to get answers that perhaps that database was not designed to [give]," said Saffell. "In other words, to find synergy using a piece of middleware that we developed."[41] It also developed Marine Air Ground Task Force (MAGTF), a tactical warfare training simulation software for the U.S. Marine Corps, which improved the Marines' operational effectiveness, readiness, and decision-making abilities, making them better prepared for warfare.

Science, Technology and Products

Whereas the Engineering and Information Services segment of Titan Systems focused on providing services to the DoD and other government agencies, the other segment of Titan Systems—the Science, Technology and Products segment—focused on more tangible products and offerings. The segment entailed "a whole different business model from the services segment," explained Ron Gorda, president of Titan Systems' Information Products Group. "You need to have engineers who can design from a blank sheet of paper a set of requirements that results in a very cost-effective, feature-rich product for the end user."[42]

The Science, Technology and Products segment was further divided into three groups—Diversified Applications, Applied Technologies, and Information Products.

Diversified Applications

The Diversified Applications Group was considered the macrocosm of Titan Systems and accounted for many of Titan Systems' technological achievements. "Most of the operations in Diversified Applications have future importance from the standpoint of technology development and product," said Bud Baird, who, in addition to his duties as president and CEO of Titan Systems, also served as group president for Diversified Applications.[43]

As of February 2002, Diversified Applications comprised two divisions. The Aerospace Electronic Division of Titan Systems provided the company with a unique cadre of advanced sensor and communications technology expertise under the leadership of Dr. Robert S. Cooper, former director of NASA's Goddard Space Flight Center and former director of the Defense Advanced Research Projects Agency (DARPA). Aerospace Electronics explored a whole new universe of subsystems and products that had direct commercial applications for the 21st century in medicine, communications, public safety, the environment, and almost every other field of human endeavor.

The Advanced Products & Design Division developed electronic products for real-time and embedded computing. It also worked with the U.S. Navy to develop a technique that extended the life of some of the older military weapon and communications systems.

Applied Technologies

Titan Systems' Applied Technologies Group (ATG), with headquarters in Reston, Virginia, was led by group president Tony Frederickson. Its five divisions provided premier technologies and technical services to intelligence and defense sectors.

The Research and Technology Division developed computer models to predict and analyze weapons effects and conducted training to support military operations. The division had been a part of Titan since 1987 and was staffed by world-renowned engineers and scientists who originally had focused on nuclear weapons effects. "They have long-standing expertise in that area," Frederickson said, "and with the end of the Cold War, they've developed expertise in computer methods for analysis of mechanical effects such as shock wave propagation and cratering mechanics from explosives."[44]

The division's unique expertise could be applied to such areas as conventional weapon effects, terrorist weapon blast effects, weapons targeting, and air strike targeting. As Frederickson explained, "We answer such questions as 'How do you take out that building?' 'How do you take out that tunnel?' 'What's the effect?' 'What weapon should I use?' 'Where should I aim it?'"[45]

The division performed forensic blast analysis of the USS Cole and World Trade Center bombings, among others. "We reverse engineer what happened based on the observed damage and usually the known approximate location of the source," explained Frederickson. By examining the size of the source, for example, and any residue left behind, the scientists could narrow down the type of explosive used, which would in turn help identify the offender.[46]

After the September 11, 2001, terrorist attacks on the United States, the Research and Technology Division analyzed prospective terrorist scenarios, looked for blast vulnerabilities in such structures as government facilities and national landmarks, and recommended where barriers should be placed and what public boundaries should be established.[47]

In addition, the Research and Technology Division's software initiatives segment enhanced

Titan Systems' Applied Technologies Group assisted the U.S. Navy's investigation of the October 2000 terrorist bombing of the USS Cole by performing forensic blast analysis.

its modeling and analysis capabilities and created cost-effective computer-based training, Web sites, Web database applications, and scientific computer applications.

The DoD used the division's software to understand the effects of biological warfare agents, for example. "We don't possess offensive capabilities," Frederickson explained. "The interest here is really understanding our vulnerabilities to an attacker's chemical or biological capabilities and, more importantly, applying these software tools to understanding the collateral consequences of a contemplated U.S. air strike on an adversary's biological warfare abilities. We want to understand the dispersal risk and hazard of striking a weapon of mass destruction because we don't want to kill noncombatants."[48]

Pulse Sciences, another division within the Applied Technologies Group, provided pulse power systems and services to U.S. defense customers as well as some international laboratories. "For several decades, Pulse Sciences has been involved in the research, design, and fabrication of large-scale systems for storing electrical energy and rapidly discharging it in a very precise way," said Frederickson.[49]

One of the division's military applications was to simulate the effects of nuclear weapons by generating short, very high power bursts of radiation in a laboratory. "We use the simulation to test military equipment to make sure it will operate under these kinds of environments," explained Phil Spence, operations manager of Pulse Sciences.[50] For example, if a reentry vehicle were exposed to x-rays from a nearby nuclear burst, Pulse Sciences' analysis helped determine whether it could survive to complete its mission.[51]

More recently, the Department of Energy used Pulse Sciences' technology in its research with lasers and other particle accelerators. A Pulse Sciences laser system was also used by the Naval Research Laboratory to help with fusion research. "Our pulse power system can operate continuously at very high power for long periods of time, with low maintenance, which is a requirement if we're ever going to achieve fusion-based power sources," said Frederickson.[52]

The Systems Engineering Division often worked side by side with its intelligence, DoD, and commercial customers to provide systems engineering, integration, software development, and training and client services. The division used Titan-engineered software and hardware, as well as COTS technology, to provide information systems, especially on systems that required handling of sensitive data. Systems Engineering also helped the Federal Emergency Management Agency coordinate local federal authorities when responding to disasters.

The Systems Integration Division designed and provided turnkey systems, hardware, and integrated software for sensitive data management, primarily for NATO's information systems. Through the division's Battlefield Intelligence Collection Capability (BICC), all NATO member nations could share their intelligence. If NATO deployed a peacekeeping mission, for example, BICC managed the required material, food supplies, petroleum, and personnel.[53]

Finally, the Information Security Systems Division (ISSD), formerly Pulse Engineering, provided specialized information security and signals intelligence systems and services, especially enterprise applications, to the intelligence community, mainly the National Security Advisory. ISSD developed a number of cryptological systems components to protect communication. It also provided engineering services and operated signal intelligence sites for the U.S. government.[54]

Information Products

Titan Systems' Information Products Group, led by group president Ron Gorda, specialized in developing, manufacturing, and supporting products for communications and Signal Intelligence (SIGINT). The group comprised six divisions.

The Communications Products Division, formerly Linkabit, continued developing its DAMA technology for advanced satellite communications products. Its cornerstone product, the Mini-DAMA,

CHAPTER NINE: TITAN TO THE RESCUE

was the backbone of UHF communications for a number of governments, including the United Kingdom, Australia, Germany, Japan, and Korea. Mini-DAMA could be found on most U.S. submarines and aircraft carriers and was also used on U.S. frigates and air platforms such as the P-3 and the E-2C Hawkeye.[55]

The centerpiece of each Mini-DAMA was the VME cards, which contained 30,000 individualized components. "The VME is essentially a card format size that makes a standard circuit board so that people who design products with a VME format can interchange or provide new technology into a product," explained Gorda. "It provides the product in open architecture so that you can slide in new technology on that format card."[56]

Communications Products also researched Intelligence, Surveillance, Reconnaissance, and Sensors (ISRS) toward development of Micro Electro Mechanical Systems (MEMS) and very large scale integration technology.

Ron Jacobson, senior vice president for advanced technologies of Communications Products, explained that the division had been trying to grow into different areas. "We've been so focused in these military communications areas that we felt it would be important to start broadening our horizons," he said.[57]

Accordingly, Communications Products became involved in the government's "dual use science and technology" program, created as a cost-sharing program with the military technology industry. "The whole intent of that project is to provide a joint commercial and military technology development opportunity," explained Jacobson. "The government puts up about half of the money, and the industry puts up the rest."[58]

As part of the dual use science and technology program, Communications Products was involved in turbo coding, which essentially enhanced bandwidth capability within a communications device. "It's a new coding technology that's been evolving over the last few years," said Jacobson. "It allows you to communicate at lower signal levels and provide a better error correction capability to the transmitted signals. That will evolve into a chip set [or software] that could be downloaded into customers' hardware to allow them to incorporate this technology into their commercial products."[59]

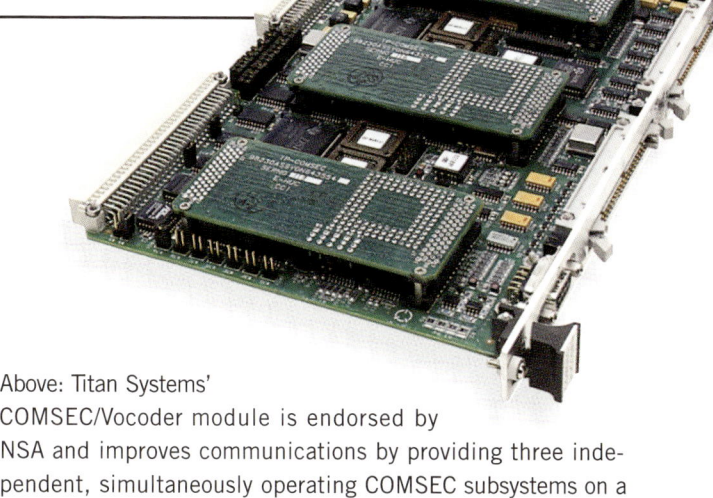

Above: Titan Systems' COMSEC/Vocoder module is endorsed by NSA and improves communications by providing three independent, simultaneously operating COMSEC subsystems on a standard VME card.

Opposite: Titan Systems' DAMA modem module consists of two independent, flexible DAMA modems and features four versatile data ports to support communication and intelligence information services.

The Signal Products Division (SPD), also part of the Information Products Group, was the world leader in small, lightweight, low-power signal intercept, direction finding, and signal processing systems for SIGINT applications. Much of its work was classified, but one of its main thrusts was in designing receivers that listened for signals. These direction finding, or DF, receivers could locate where a signal emanated from so it could be decoded or intercepted.

One of SPD's key products was called "Prophet," a new electronic signal interception and jamming system that could be placed on top of ground vehicles or helicopters. Gorda explained Prophet's DF capability:

Essentially you listen to the signals of other people communicating and then try to find out where that communication is emanating from. Not only do you intercept it and hopefully decode it, but you also then can provide what's called electronic attack, where once you find where the source of the communication is, you can electronically jam it and prevent an adversary from communicating.[60]

To win the Prophet contract from the U.S. Army, multiple divisions of Titan Systems worked

Above: Titan Systems developed Prophet Ground for the U.S. Army. Prophet is an electronic signal interception and jamming system that can be mounted on ground vehicles such as Humvees or on helicopters.

Right: Datron's Manpack radio became the "radio of choice" for defense organizations around the world.

together on the SPD-led project. In addition to providing 83 Prophet electronic signal intercept systems for the Army, Titan Systems supplied training, fielding, and testing support.[61] In 2001 and 2002, Prophet was highly effective in helping locate terrorist al Qaeda forces in Afghanistan and the Taliban regime harboring them.[62]

After Titan acquired Datron Systems, two new divisions were created within the Information Products Group. The Datron Advanced Technologies Division designed, developed, and manufactured antennas and imaging system products. Its communication solutions included remote-sensing satellite ground stations; mobile broadband communication systems for airlines and military transports; and mobile air, land, and marine direct-broadcast satellite television systems.

The division's large-antenna TT&C (telemetry, tracking, and control) systems were capable of tracking such things as a meteorological satellite. "What we do is track the satellite and actually take a download of weather data from that satellite," said Gorda. "We process that data and provide real-time weather information."[63]

In addition, Datron Advanced Technologies was busy developing a broadband antenna system to send high-data rate information to a mobile platform. The antenna system could provide Internet access to an airplane, for example.

The other division elaborated from Datron Systems, Datron World Communications, manufactured handheld and mobile communication devices for military and government agencies, mainly in developing countries. Its Manpack radios and mobile radios were installed in vehicles and provided both ultra-high frequency (UHF) and very high frequency (VHF) communication.

The division's APCO Project 25 radio was tailored more for the civil agency marketplace than

the military. It provided a digital radio standard that would be especially helpful in times of emergency. Police, fire, and other emergency services all used radios, whether handheld or installed in vehicles. "A lot of these emergency agencies have to interoperate and communicate," explained Gorda, "and without a solid standard, they cannot talk efficiently."[64]

The Systems and Imagery Division, formerly DBA Systems, which Titan had acquired in 1998, continued its work in image processing products, electro-optics, and integration capability. Its fingerprint scanners were an important addition to the FBI's capabilities.

Finally, the Communications and Software Solutions Division provided unique solutions to satellite communications challenges. NASA employed its OpenGL display graphics package and dynamic simulation system to train astronauts and ground control support personnel.

Cayenta

While Titan Systems gave Titan Corporation a solid base from which to grow, other Titan subsidiaries faced difficult market environments. Cayenta, the company's e-business solutions subsidiary, had grown substantially since its official formation in 1999. As early as March 2000, an analyst with B. Riley & Company predicted that Cayenta would "be a home run," partially because it had "an experienced team that has been together for a long time."[65]

Under Dave Porreca's leadership, Cayenta grew by 69 percent during 2000, building, hosting, and managing complex, end-to-end, traditional and Internet-based information management systems. Cayenta provided software for finance and accounting, customer billing and collection, multichannel marketing, and enterprise asset management. It implemented and integrated complete technology solutions and provided ongoing business services management. Its software solutions were especially helpful to utility and transportation industries that had very complex e-business transactions.

Early in 2000, Cayenta introduced Utility-Manager™ 7.0. Designed specifically for the utility industry, this software provided electronic bill processing, multiservice billing, flexible rate

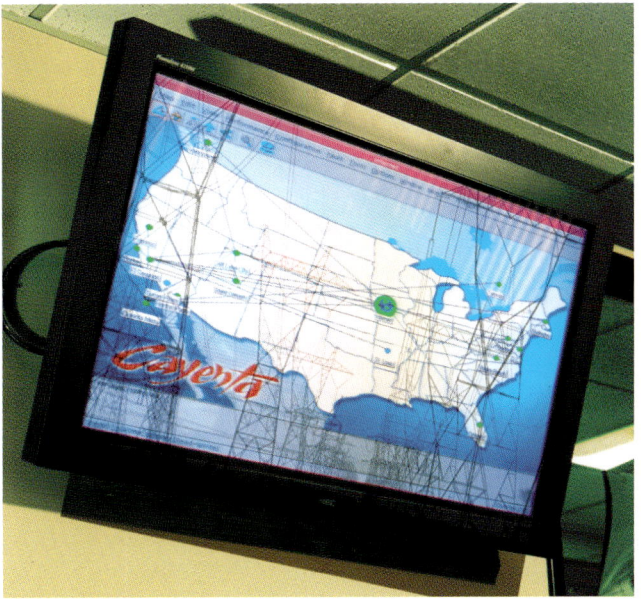

Cayenta provided a full complement of Internet-based management services. The company had several regional solution centers around the country and in Canada that were used to manage and monitor hosted operations.

structures, and on-line enrollment and payment capabilities. It was a hit with municipalities. Among those who became users were Orange County, Florida; Winston-Salem, North Carolina; and Brownsville, Texas.

MainsaverNow™, a hosted version of Utility-Manager, caught the attention of Florida Power & Light, one of the nation's fastest-growing utilities. After the Florida utility signed on as a customer, so did NRG, a power generation facility for northern states.

Mainsaver was so efficient in helping to "quell the chaos of modern maintenance operations" that it was recognized by *Plant Services* magazine in 2001. The magazine's *MRO Marketplace* supplement selected Cayenta's Mainsaver Enterprise Asset Management (EAM) as a Gold Winner of its annual "MRO Problem Solvers Award" competition. Mainsaver EAM won for saving the Chicago water district over $2.25 million a year.[66]

Cayenta e.System became the first direct marketing system to use standard Internet browser technology. It was also the first direct marketing system that allowed merchandisers to track their customer activity, orders, inventory, and supply-chain activity with a single system. After reviewing more than 50 different e-business solutions, in December 2000 Express.com, a leading entertainment e-commerce retailer, chose Cayenta e.System to support its Internet operations, calling it "hands down the best out there."[67]

In 2001, Cayenta ran into the same brick wall that many other commercial information technology businesses faced in the down economy. Revenue dropped from $84 million in 2000 to below $60 million, and the company was forced to lay off about 200 people. Fortunately, Cayenta's vertical market focus—targeting utilities and municipalities, manufacturers, and retailers—helped it withstand the worst of the market conditions, at least for awhile.

Though Cayenta's corporate customers spent less on information technology products and services during 2001, Cayenta won a number of new customer contracts that year, especially from municipalities. In April, the FAA's Office of Airport Planning and Programming selected Cayenta to modernize two of its key information systems. In June, Cornerstone Promotions, the leading direct-TV advertising agency, chose Cayenta Assist™ to improve its customer database management capabilities. The next month, Cayenta partnered with ElectriCities, a trade organization representing public power communities, to offer ElectriCities' member cities a new customer information system that helped streamline billing procedures and increase efficiency. Then in September, Sithe Energies, one of the world's top five independent power producers,

As a Total Solution Provider, Cayenta built, hosted, and managed complex, end-to-end, traditional and Internet-based information management systems.

Cayenta combined proprietary software with the implementation and integration services needed to deliver complete information technology solutions to customers.

selected Cayenta's Mainsaver EAM solution for 14 plant upgrades.

During 2001, Cayenta produced enhanced versions of UtilityManager, MunicipalityManager, and Mainsaver and released two more software services in January 2002. Cayenta Assist Software Services was designed to bring a cost-effective, fully capable direct-commerce management system to retailers of all sizes. Cayenta WebStore™ software allowed retailers to easily and quickly implement a Web-based commerce management system.

Dave Porreca retired in March 2002 but remained available to continue consulting with and advising Titan. Porreca had left a major imprint on Titan during both of his tenures with the company, the first from 1982 to 1989 and the second beginning in 1999 when Titan purchased Transnational Partners, the company Porreca had founded.

Unfortunately, by that time, it had become clear that the commercial information technology market was not going to rebound in the near future, and Titan made the strategic decision to sell certain of Cayenta's operations in order to focus on the most promising growth opportunity: the utility and municipality market, where Cayenta's products enjoyed leading market positions.

Cayenta's utility and municipality offering continued to expand its customer reach and brought cost-effective revenue cycle management solutions to utilities and municipalities in the United States and abroad. Most importantly, the company continued to be an important source of profit for the Titan Corporation.

Titan Wireless

Titan Wireless experienced an astounding 198 percent increase in revenue in 2000 as it continued to bring low-cost satellite communication solutions to developing countries around the world.

Titan Wireless' strong performance was based on its ability to develop a low-cost structure, to find the right local partner, and to measure and manage the risk of doing business in developing countries. Titan Wireless capitalized on the opportunity to provide telecommunications services to underserved markets. "What others may see as challenges, Titan Wireless sees as opportunities," the company reported.[68]

Besides adding new gateways to its network, Titan Wireless focused on developing a corporate retail business during 2000. That business provided voice, data, and broadband connections through a combination of very small aperture terminals (VSATs) and terrestrial-based wireless, local-loop technology.

In addition, Titan Wireless launched telecom service in Benin, West Africa, with a nine-year contract. Titan installed and operated components of the government's telecommunications infrastructure, including a GSM cellular network, a rural telephony network, a fiber-optic backbone, and local telephone switching equipment. By the end of 2001,

Above: Titan Wireless' satellite communications systems and services provided low-cost telecommunications services, mainly to underserved areas.

Below: Titan Wireless installed this satellite dish in Benin, Africa, in 2000.

This was Titan Wireless' first foray into the retail service business.[70]

Also in 2001, Titan Wireless completed a cellular system in Guinea for a U.S. company and a satellite backhaul system for a major fixed wireless carrier in Nigeria.[71]

By 2002, the deterioration in the international telecommunications market had begun to affect Titan Wireless' business as operators and foreign governments in developing countries were no longer able to obtain financing for the build out of networks. This significantly impacted Titan Wireless' systems integration business.

By mid-year, Titan's long distance business had also become problematic as well. Certain competitors, by this time operating in bankruptcy or undergoing significant reorganizations, sought to generate revenue by reducing pricing, which created significant margin pressure.

As a result of these developments, Titan Wireless could no longer generate the significant operating profits it had once enjoyed. Because Titan wanted to focus management resources on opportunities in its growing core defense business, in July 2002 Titan made the decision to

the GSM mobile network had attracted more than 70,000 subscribers. Such rapid popularity confirmed that demand for mobile telecommunications service was high in developing countries.

Despite the slow economy and a deteriorating telecom market in 2001, Titan Wireless maintained its steady growth by continuing to focus on underserved markets. By January 2001, Titan Wireless had installed nearly 40 satellite gateways in underserved markets in Africa, the Middle East, and Latin America.[69]

In February 2001, Titan expanded wireless further, purchasing a minority interest in Contech Media, a new company in the newly liberalized telephone market of Estonia. Five months later it acquired a minority interest in Gateway System, which provided Internet service to India's corporate market. Through Titan Wireless' Sakon subsidiary, Gateway began marketing Titan Broadband, which provided broadband data and Internet services to corporate customers in India including banks, hotels, and the Chittagong Stock Exchange.

This map highlights all the countries where Titan Wireless provided communications services and systems. The company focused on developing countries, primarily in Asia, Africa, the Middle East, and Latin America.

eliminate its exposure to the telecommunications market. While the decision was difficult, the company knew that it was in the best interest of its shareholders to exit a noncore business that was in an industry that showed no sign of recovery in the near future.

Emerging Technologies

As part of its continuing strategy to create new commercial businesses from existing technologies, in 2000 Titan developed three new subsidiaries under its Emerging Technologies and Businesses segment: E-tenna and LinCom Wireless (both of which came from Titan Systems), and Titan Scan (which had been joined with SureBeam). Each subsidiary received its own new leadership and an employee stock ownership plan.

E-tenna developed and licensed unique radio frequency and electromagnetic technologies for the wireless market. One of E-tenna's major products was an embedded antenna for smaller, sleeker cellphone designs. By teaming with outside investors, it had established a new global antenna standard by using its Artificial Magnetic Conductor (AMC) technology. AMCs are thin, low-cost printed circuit boards with unique and powerful electromagnetic properties that allow wireless antennas to be isolated from nearby influences that could degrade

E-tenna, one of the companies in Titan's Emerging Technologies and Businesses segment, focused on the mobile wireless market. E-tenna scientists like these developed embedded antennas for smaller, sleeker cell phone designs.

performance. Because AMCs are so small, antennas could be lighter, smaller, and more flexible than with previous technologies.

Another E-tenna product, called AccuWave, employed the AMC technology and was used in global positioning systems (GPS). AccuWave was designed to improve the accuracy of GPS systems while also reducing the size of their antennas. Improved accuracy was especially important for high-precision applications such as surveying.[72]

In May 2001, E-tenna opened its new testing and research development facility in Laurel, Maryland, to develop and demonstrate the benefits of its technology. The facility included the United States' first Satimo test chamber, which allowed for the most accurate and rapid measurements of antenna performance available.[73] A strategic alliance with Ashvattha Semiconductor to combine E-tenna's reconfigurable antenna technologies with Ashvattha's RF chip technologies also promised to significantly advance the wireless market.[74]

To lead this growing company into its next phase of development, industry veteran Steven Grossman was named E-tenna's president and CEO in July 2002, giving the company sales and marketing expertise to complement its excellent technical research team.

LinCom Wireless was led by Patrick Henry, who joined Titan in October 2001 after spending many years in the semiconductor and home media

FRIENDS IN CONGRESS

THE PRAISE FOR GENE RAY HAS NOT been limited to Titan's people. Political leaders in Titan's home state of California have been especially appreciative of the benefits Titan has brought to their state and the entire country.

"Gene Ray has used technology to improve the lives of Americans," said Congressman Randy "Duke" Cunningham, who represented California's 51st Congressional District. "Through his initiative, Titan has been able to do wonders. Our personal security is enhanced through his vision, and our national and economic security as well." This was high praise indeed considering Cunningham's patriotic track record. He was one of the most highly decorated pilots of the Vietnam War and was the war's first fighter ace. He also trained fighter pilots at the Navy Fighter Weapons School in the "Top Gun" program at Miramar Naval Air Station.

Congressman Cunningham pointed out that Titan provided "a lot of jobs, and Ray's innovation has enabled Titan to grow the number of jobs. President Bush's economic stimulus package on investing in business to create jobs is so important, and Gene Ray is a part of that."[1]

Congressman Duncan Hunter, of California's 52nd Congressional District, complimented Ray for his "extraordinary vision and leadership capability" and observed that companies like Titan that dealt in both commercial and defense markets have a "superior perspective" to those companies who deal solely in the defense sector. Before he became a congressman, Hunter had his own law practice and served in the 173rd Airborne and 75th Army Rangers during the Vietnam War. As chairman of the House Military Research & Development Subcommittee, he oversaw development and testing of military systems, weapons programs, and military-related technologies.

"The art of the possible is much more obvious because of corporate leaders who have a hand in both commercial and military arenas," Congressman Hunter said. "Gene has a hand in both arenas. Also, I think one of his greatest gifts is his ability to motivate and inspire people and to focus on the goals that ultimately put a product on the shelf that enhances American security."[2]

industries. LinCom's former CEO and founder, William Lindsey, became LinCom's chairman.

LinCom's products promised to bridge the gap between the dominant Wireless Local Area Networking (WLAN) standard 802.11b (which operates in the 2.4 GHz frequency) and the next-generation, higher speed 802.11a (which operates on a less congested 5 GHz frequency) by cost-effectively supporting both modes of operation. The pervasiveness of 802.11b access points and the anticipated adoption of 802.11a access points as a dominant network standard meant LinCom's combination 802.11a/802.11b chipsets and board-level products would be in high demand. And because LinCom's solutions were cost effective, they would likely spur rapid and widespread development of wireless multimedia devices in homes, small businesses, and enterprise settings. The WLAN chipsets that LinCom was developing, for example, promised to provide full wireless functionality for the home, including wireless television.

Titan Scan was the fourth subsidiary in the Emerging Technologies and Businesses segment and had already had its own spin-off from the SureBeam® food irradiation technology. Operating from three sterilization facilities, Titan Scan used SureBeam to sterilize a long list of medical products, ranging from disposable devices to petri dishes and surgical gowns and gloves. It also sold its systems worldwide to medical product manufacturing companies. In 2001, Titan Scan began marketing its electron beam technology to improve material performance in semiconductors, aerospace composites, wire and cable, plastics, and packaging films.[75]

SureBeam Corporation

The enormous potential for irradiated foods had become clear in the late 1990s, and Gene Ray knew he needed to bring in talent to develop that potential to its fullest. Thus in November 1999, Larry Oberkfell, who had experience running food businesses, came aboard as a Titan senior vice president to head up the electronic sterilization and pasteurization effort.

Before Oberkfell joined the SureBeam team, considerable investments and major progress

Larry Oberkfell is president and CEO of SureBeam Corporation, which successfully completed its IPO in March 2001.

had been made to get the business started. Titan had invested about $10 million to build the world's first electron-beam food irradiation facility in Sioux City. Though the facility had been irradiating food for test marketing, government approvals had not yet been issued for the processed and ready-to-eat food markets. SureBeam had also signed exclusive agreements with the world's largest beef and poultry producers—Cargil, Tyson, and IBP—to test SureBeam technology.

The month after Oberkfell was hired, in December 1999, the USDA gave its initial okay to use electron-beam technology to kill pathogens in meat. The following February, government regulations became official.

Kraft Foods saw the potential for SureBeam's technology, and in February 2000 Titan and Kraft formed an agreement to work jointly on research surrounding the use of technology to treat ready-to-eat foods. Though the government had yet to approve the use of electron-beam technology for use on ready-to-eat foods such as lunch meats and hotdogs, the partnership agreement was a sign that Kraft felt the approval was only a matter of time.

In May 2000, three months after the USDA published its official regulations governing irradiated red meats, Titan brought its first SureBeam-treated food to the retail market with Minnesota-based Huisken Meats' beef patties.

By then, Titan had formed SureBeam Corporation for the purpose of irradiating pathogens in food; Titan Scan remained a separate entity.

SureBeam acquired certain assets of Applied Power Associates (APA), a leading engineering and architectural firm that specialized in testing and facility design for the integration of electronic pasteurization systems into food and meat processing plants. APA had been a leader in advocating and adapting the use of electron-beam technology for meat and other types of food.

Also in May, SureBeam formed a joint venture with Brazil-based Tech Ion Industrial to build a network of SureBeam facilities in that country. Titan would own a minority interest in the joint venture, which would be called SureBeam Brazil.

The SureBeam business was moving rapidly, but Ray and Oberkfell wanted more than that; they wanted the company to be the leader in e-beam technology. Toward that goal, SureBeam formed an alliance with Texas A&M University in June 2000 for continued research on electron-beam pasteurization technology.

Being a leader in the business also meant patenting the technology and protecting Titan's intellectual property. By the end of 2001, SureBeam had been awarded its 13th U.S. patent. All of the patents further advanced SureBeam's leading position in electronic irradiation technology.[76]

A Growing Industry

Despite opposition from such organizations as Public Citizen, the SureBeam® technique was becoming more accepted as the public became better educated. In Minnesota, for example, where Huisken's SureBeam beef patties were sold, an October 2001 poll indicated that 77 percent of consumers said they felt safe eating the product.[77]

"After the government recognition and after people started to realize that the aroma, taste, and appearance of the food weren't altered as they had been under the old technology, people began to accept what we were doing," said Gary Loda, adding that there was probably less opposition to irradiation of food than there had been to pasteurization of milk.[78]

Or, as Oberkfell pointed out, "We were able to demonstrate that people's concerns were unfounded by presenting proof. Eating is proof. You eat it; you believe it."[79] Oberkfell told the *San Diego Union-Tribune* that acceptance of irradiated food increased to more than 80 percent after consumers were educated about its benefits.[80]

In May 2000, Huisken Meats introduced SureBeam-processed beef patties to the public. By the end of the year, SureBeam beef patties could be found all over the nation under many different brands.

Some debate persisted over whether SureBeam should be used to eliminate pathogens in ready-to-eat and processed meats. *E. coli* and salmonella are found in raw foods, but listeria bacteria are found in precooked, ready-to-eat foods. Though the FDA and USDA had yet to approve the use of SureBeam® technology on precooked foods, SureBeam's future looked even more promising when, in the spring of 2000, President Bill Clinton called for regular listeria testing of hot dogs, luncheon meats, and other prepared meats.

However, SureBeam's most important moment of acceptance probably came when Huisken, which later became a unit of Sara Lee, introduced its first SureBeam®-processed beef patties in 84 markets in May 2000. As consumer demand grew, so did the availability of SureBeam-treated beef patties. By year's end, the patties would be sold in some 2,000 markets across the country. And by the end of 2001, SureBeam-processed ground beef could be found in thousands of supermarkets located in over two-thirds of the nation.

Adding to that success, Schwan's, a leading home distributor of premium frozen foods, Omaha Steaks International, renowned for its gourmet meats, and WW Johnson Meat Company, a prestigious national processor, began selling SureBeam-processed ground beef. In addition to its partnership with Kraft, the largest packaged-food company in North America, SureBeam formed partnerships with other ready-to-eat food companies: Del Monte, Anchor Foods, and SCIS Food Services.

In November 2000, Titan signed an agreement to begin installing a SureBeam in-line system at a meat-processing plant owned by IBP, the world's largest meat-processing company. Under SureBeam's in-line system strategy, IBP would pay the company a fee for every pound of meat treated by SureBeam. Later, it installed in-line SureBeam systems in two Excel processing plants. Excel was the nation's second-largest meat company.

By the end of 2000, the remarkable results were in: SureBeam had increased revenue an astounding 573 percent over 1999's to $25.2 million. But more importantly, SureBeam had revolutionized the food industry.

SureBeam Spins into Orbit

With such success and such a grand future awaiting, SureBeam Corporation was ready for its IPO, and on March 16, 2001, SureBeam was launched as a publicly traded company on Nasdaq, with Titan maintaining an 84 percent interest in the spin-out. Though the IPO faced a down market, SureBeam was able to sell 6.7 million shares at $10 a share, which met Titan's expectations. Titan planned to spin off the remaining shares to existing Titan shareholders as a tax-free dividend.

"It was the scope of our technology that sold our IPO," said Oberkfell. "If you take red meat, poultry, ready-to-eat meat, fruits and vegetables, and seafood—just those five categories, just in the United States—and if we have only a 10 percent niche in treating those products, the sales revenue of this company would be $1 billion."[81]

A SureBeam facility in Chicago opened in 2001 and another in Los Angeles in 2002. All the facilities were capable of using the electron-beam and x-ray technologies simultaneously to accommodate different product sizes and shapes. "So whether we want to treat a hamburger patty or a whole turkey, the facilities will be able to do it," explained Oberkfell.[82]

Also in 2001, RESAL Saudi Corporation, a subsidiary of a private Saudi Arabian conglomerate in Riyadh, agreed to build multiple SureBeam facilities throughout the country for pathogen and environmental pest control.[83]

New Uses for SureBeam

By December 2001, Gene Ray was getting used to bright lights and reporters' questions. His once "little-known" company was now in the national spotlight. *ABC News*, *USA Today*, and *Bloomberg News* all carried features on Titan, for once again

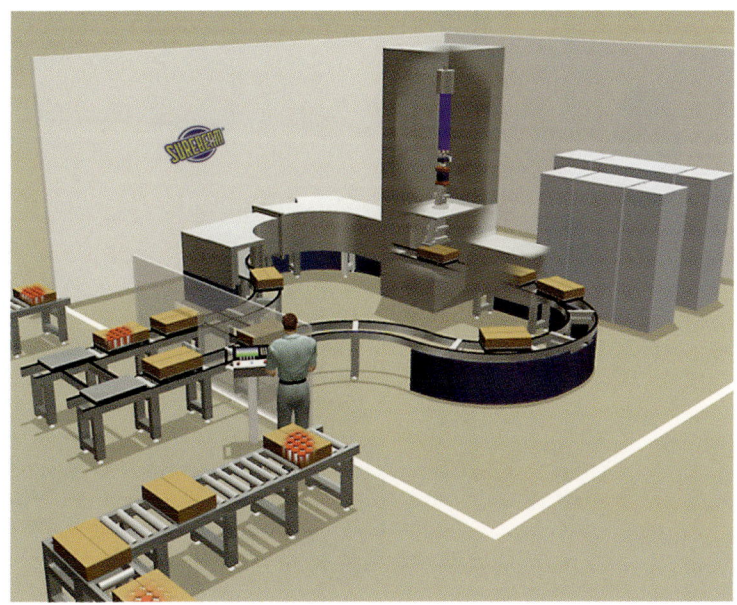

Titan's SureBeam in-line systems are slated to be installed in an IBP-owned meat-processing plant and in two Excel processing plants. IBP is the world's largest meat-processing company and Excel is the second-largest meat company in the United States.

Titan had taken an old defense technology and found a new—and important—use for it. This time, Titan's SureBeam® units were being sold to the U.S. Postal Service to irradiate mail.

The tragic September 11 terrorist attacks on the World Trade Center and the Pentagon had been followed by another attack: anthrax-tainted letters mailed to high-ranking public officials and media personalities. It was a mind-boggling problem, but Titan had a solution: irradiate letters, kill anthrax. In a startling instance of foresight, Titan—always on the lookout for new uses for its technologies—had issued a report in July 2001 saying that its electron-beam technology could be used for that very purpose. "Any postal delivery service could potentially be used to deliver biologic weapons," Loda observed, "and our process doesn't care if the microbes are on food or on envelopes."[84]

In October 2001, Titan created a Homeland Security Office headed by former San Diego Mayor Susan Golding. The office would provide government and commercial customers with integrated solutions

Above: The U.S. Postal Service purchased eight turnkey SureBeam systems to help ensure that mail was safe to handle.

Left: The anthrax postal threat prompted the United States Postal Service to produce and display "suspicious mail alert" posters.

to the bioterrorism threat.[85] Its first move was to offer SureBeam's technology to eliminate anthrax contamination from the U.S. mail, and almost immediately the U.S. Postal Service contracted with Titan. In October a Titan Scan medical supply sterilization facility in Lima, Ohio, began sterilizing mail from the anthrax-tainted facility in Washington, D.C. In addition, the Postal Service purchased eight complete turnkey systems to irradiate mail.[86]

With tons of suspected contaminated mail en route to Washington, New York, and other places being shipped to Titan for sterilization, the nation's media focused on Titan and its technology. For a two-week period starting in late October 2001, Titan received over 400 media queries as the press sought information on how the company's electron beam technology was eliminating the threat of anthrax in the mail. The demand for information was fierce. On one morning alone, Ray completed eight interviews before 8:00 A.M. in order to accommodate the early East Coast morning show broadcasts. By the end of November, over 1,000 stories had run on Titan's technology, and Ray had done interviews with virtually every national news wire, newspaper, and magazine, as well as all of the prominent television news programs.

"Titan . . . will be largely responsible for safeguarding mail that crosses the desks of the nation's leaders," reported USA Today in a cover story titled, "Titan Zaps into the Spotlight." In interviews with ABC, NBC, FOX, CNN, PBS, and CBS, Ray, using easy-to-understand terms, explained how Titan's very sophisticated technology worked. "I actually got it," said CNN's news anchor Paula Zahn. "[He] made it very easy to understand." As it turned out, Zahn's remark underscored an achievement no one had contemplated.[87]

A Porter Novelli survey of U.S. adults following the news coverage of Titan found that consumers had moved to a strong level of support for irradiation technology—with more than half (52 percent) said the government should require irradiation to help ensure a safe food supply. Titan's effort to build an understanding of how irradiation could be effectively used to stop bioterrorism in the mail had also resulted in a dramatic shift in public opinion on irradiated foods.[88]

The SureBeam® technology could be used against other bioterrorist threats as well. During the Gulf War, the United States had attempted to destroy Iraq's supplies of nerve gas and other biological weapons supplies by burning or exploding them. But as Loda explained, "That's not a terribly good way to try to destroy microbes. You can oftentimes spread them."[89]

Since the 2001 terrorist attacks, SureBeam had been in talks with the military about the possibility of using the basic technology to decontaminate offensive weapons in the field. "After the battle is over, you set up the machine, and it picks up whatever munitions might have been designed to deliver these agents and activate them," Loda said. "If there is a biologic event in a conflict, obviously there's going to be a lot of material that is contaminated. Uniforms, for example, or the suits soldiers put on to protect themselves against biologic attacks. After the conflict is over, you have to do something with those suits. We have had discussions with people about taking those suits and passing them through one of our systems as a way of cleaning them so they can be reused."[90]

Even in a nonbiological conflict, SureBeam could be used to destroy harmful pathogens on soldiers' food. "So many soldiers are not able to perform because they get sick from eating the local foods," Loda said. During the Gulf War, for example, some pilots were sidelined because they grew sick from the food they ate.[91]

SureBeam technology can be used to decontaminate nuclear-biological-chemical suits so they can be safely stored and reused.

Acknowledging the benefits of the SureBeam® technology, in November the DoD added irradiated ground beef and poultry products to its normal military food procurement lists.

More Acquisitions

Titan continued growing in 2002. In March, it acquired Jaycor, a San Diego–based maker of sensors, radar, and communications electronics for military customers. Like The Titan Corporation, Jaycor began as a defense contractor, in 1975, and later created commercial businesses from its DoD technology. Jaycor's key products included respiration sensors, acoustic imaging sensors, visibility sensors, radar systems, radiation-hardened and ruggedized military devices, and network security products. The sale added about $45 million to Titan's 2002 revenue and reduced its debt by $50 million. Jaycor became a subsidiary of Titan Systems and continued to be run by Eric Wenaas, Jaycor's chairman and CEO.[92]

On the same day as the Jaycor deal closed, Titan completed another deal, buying GlobalNet, Inc., which became a part of Titan Wireless. Founded in 1996, GlobalNet was a leading Voice-over-Internet Protocol carrier that specialized in providing international long-distance service to U.S. and foreign-based service providers operating in Latin America. DeMarco noted that the purchase would nearly double the size of Titan Wireless' revenue base, increase its presence in Latin and America, and expand its global network.[93]

Earlier in the year, Titan had announced it would purchase Science & Engineering Associates (SEA), an employee-owned company based in Albuquerque, New Mexico. SEA was founded in 1980 as a defense research and development contractor. Over the years, it moved into systems integration, modeling and simulation, and custom software design services for government and military customers.[94]

Fostering Innovation

Those who played pivotal roles in Titan's ongoing success were rewarded with promotions that would allow them to better carry out their proven leadership. Eric DeMarco, previously executive vice president and chief financial officer, was promoted in April 2001 to chief operating officer and to president and COO of The Titan Corporation in the spring of 2002. Mark W. Sopp, meanwhile, was elected senior vice president and CFO. Sopp had joined Titan in 1998 as vice president and CFO of Titan Systems.

Titan's management team appreciated the importance of each of its more than 11,000 employees and worked hard to create a corporate culture and work environment that fostered innovation, high ethics, and employee satisfaction. "We are mindful that our most valuable asset has been and continues to be our employees," the company reported. "We are proud that they have chosen to work at Titan and are mindful of the fact that they are the reason for our success."[95]

Eric DeMarco joined The Titan Corporation as senior vice president and chief financial officer in January 1997. He became executive vice president in 1998 and was promoted to chief operating officer in 2001. DeMarco was promoted to president and COO of The Titan Corporation in 2002.

"It's very important to me that our employees be proud to work for our company," said DeMarco. "We have done some things that may seem trivial, but I believe they are very important in building the team so that people are motivated to come here."[96] The "trivial" things that DeMarco was referring to involved adding a full gymnasium to Titan's headquarters, holding raffles, and giving away sporting events tickets. They came on top of various other rewards and incentives and, of course, employee stock ownership plans.

All over the corporation, whether at San Diego headquarters or at its hundreds of offices around the world, employees enjoyed remarkable flexibility and the freedom to create, to innovate, and to share their ideas. It was a corporate culture that came from the top.

"The corporate office doesn't micromanage the individual divisions," said Phil Spence. He had been president of PSI when Titan acquired it in 1987 and was speaking from personal experience. "They bring in people they have confidence in and let them thrive."[97]

"Gene Ray is a participatory manager," said Bill Zettinger, vice president of programs at Titan Communications Products. "He leads by example, and the values that he hands down—respect for the individual, taking care of customers, not closing your technical mind—are so important to the corporation. The Titan culture allows for freedom of expression, and it's a culture that gives you the responsibility for getting things done."[98]

On November 7, 2001, the New York Stock Exchange (NYSE) honored Titan by inviting Gene Ray and the Titan board to ring the bell that triggers the stock exchange's opening. Front row, from left: Daniel Fink, Joseph Caligiuri, the NYSE chairman, Gene Ray, Susan Golding, and Robert Hanisee. Back row, from left: Charles Allen, Jim Roth, Robert LaBlanc, Joseph Wright, Eric DeMarco, an NYSE representative, and Michael Alexander.

Primed for the Future

As The Titan Corporation approached its 22nd anniversary in 2003, there was no doubt that it was well positioned for future success. High-technology corporations like Titan could only benefit from President George W. Bush's proposed 2002 budget boosts for defense and homeland security. And Titan's ongoing innovative technological advances in communications and information systems had not only exposed the world to future benefits but also positioned the company to take advantage of future opportunities. Analysts forecast Titan's 2002 revenues to reach $1.5 billion. As *Washington Technology* magazine observed in January 2002, "The emergence of government and defense markets as a bright spot in the U.S. economy is now widely understood. . . . At the end of the day, it's a great time to be in the industry."[99]

Or, as Rochelle Bold, vice president for investor relations, pointed out, "We have become a relatively stable business in a very unstable business environment. The investment community shows a lot of interest these days in defense contractors."[100]

In the months following the September 11 terrorist attacks, Titan saw an increasing demand for Signal Products' Prophet technology as well as many other Titan products and services. It provided bioterrorism expertise, for example, and emergency management and support for the victims of terrorism in New York, Pennsylvania, and Washington, D.C. It also provided emergency preparedness support to the Department of Justice.[101] In addition, Titan had been working for years on interactive software tools to combat terrorism and new technologies to protect vital information in computer systems from being disrupted or destroyed.

"With the international situation the way it is now, our marketplace is even broader than it was before," said Gorda. "We have always provided our solutions to NATO countries. Some of our technology has classified aspects to it, where we're not allowed to sell to certain countries because we don't want to provide those kinds of capabilities to potential adversaries, but now we have a broader set of friends, countries that have joined the coalition that is in place now with the war in Afghanistan."[102]

The quality of Titan's leadership also set it apart and equipped the company for continued success. Ray observed that he had been "blessed with an outstanding board of directors," and the board itself was active in decision making.[103]

In the spring of 2002, The Titan Corporation's board of directors consisted of Gene W. Ray, chairman, president, and CEO; Michael B. Alexander, former chairman & CEO of AverStar; Dr. Edward H. Bersoff, chairman and CEO of Re-route Corporation and former chairman, president, and CEO of BTG; Joseph F. Caligiuri, former executive vice president of Litton Industries; Peter A. Cohen, founding partner of Raimius Capital Group; Daniel J. Fink, former senior vice president of corporate planning & development of General Electric Corporation; Susan Golding, the former San Diego mayor; Robert Hanisee, chief investment officer for Asset Allocation in the Private Client Services Group of Trust Company of the West; Robert E. LaBlanc, president of Robert E. LaBlanc Associates; Jim Roth, former chairman, president, and CEO of GRC International; and Joseph R. Wright Jr., vice chairman of Terremark Worldwide, president of Terremark Communications, and president and CEO of PanAmSat.

Two longtime Titan board members were testaments to the strength of Titan's leadership. "The board is made up of people who are not interested in posturing," said Dan Fink, who joined the board in 1984. "We get along very well, and there is a very open relationship between the board and Gene and the other officers. All that leads to success."[104]

Joseph Caligiuri, a Titan board member since the company merged with EMM in 1985, also felt that The Titan Corporation had exceptional leadership. "The board and Gene believe in each other, and we have worked hard together," Caligiuri said. "Gene is very open and discusses critical issues facing the company openly with the board. Eric DeMarco, too, is a brilliant young executive. He came in as CFO, but he has shown such great management skills that it soon became apparent that he was destined to be involved in the overall management of the corporation."[105]

And across the entire Titan Corporation, executives gave much of the credit for Titan's success to Gene Ray, the cofounder, chairman, president, and CEO.

"Gene's leadership is phenomenal," said Mary Squazzo, who joined Titan in 1986 and became

By founding and leading The Titan Corporation, Gene Ray, shown here with his five grandchildren, has done much to make the world a safer place for future generations.

Ray's executive assistant in 1995. "He has a can-do attitude all the way. You can't say to him, 'Sorry, Gene, we can't do that; we just don't have the resources' because he'll tell you that we'll find a way. That fighting spirit helps the company grow, and it motivates the employees."[106]

"There's been so much consolidation in our industry, starting with the defense budget cuts in the late 1980s," said Gorda. "If it weren't for Gene's strategy—if we had stayed at the $100 million to $200 million level—I think we would have been gobbled up by now."[107]

Ray himself remained humble about his achievements, always crediting Titan's employees for the company's amazing success. "We have creative people who are enthusiastic about coming up with new ideas," he said. "They come up with innovative ideas in order to do their jobs well for customers."[108]

It was a sentiment that harked back to Titan's earliest principles: excellence in all endeavors, customer satisfaction, corporate integrity, and respect for the individual. And though The Titan Corporation had certainly continued fostering these principles, its formula for success and growth had evolved to keep pace with the world's ever changing social, political, and economic conditions.

And that's what The Titan Corporation would continue to do. Its success at diversification—at creating new technologies, developing innovative technological advances, and finding new markets—enabled it to survive and thrive, no matter how uncertain the market, no matter how uncertain the world.

NOTES TO SOURCES

Chapter One

1. Gene Ray, interview by Jeffrey L. Rodengen, tape recording, 24 May 2001, Write Stuff Enterprises.
2. Ed Knauf, interview by Richard F. Hubbard, tape recording, 12 November 2001, Write Stuff Enterprises.
3. Jack McDougall, interview by Richard F. Hubbard, tape recording, 4 December 2001, Write Stuff Enterprises.
4. Anthony Ramirez, *Los Angeles Times*, 16 August 1983.
5. Knauf, interview.
6. Richard Llewellyn, interview by Richard F. Hubbard, tape recording, 8 November 2001, Write Stuff Enterprises.
7. Knauf, interview.
8. Victor Gogolak, interview by Richard F. Hubbard, tape recording, 3 December 2001, Write Stuff Enterprises.
9. Cathy Pergam-Ball, interview by Richard F. Hubbard, tape recording, 19 December 2001, Write Stuff Enterprises.
10. Ibid.
11. Knauf, interview.
12. Ray, interview, 24 May 2001.
13. Ibid.
14. Ibid.
15. Ibid.
16. Ibid.
17. Ibid.
18. Information obtained from the SAIC public relations department.
19. Ray, interview, 24 May 2001.
20. Ibid.
21. Ray, interview, 24 May 2001.
22. Knauf, interview.
23. Ibid.
24. Ray, interview, 24 May 2001.
25. Ibid.
26. Ibid.
27. Gogolak, interview.
28. Linda Frady Keenan, interview by Richard F. Hubbard, tape recording, 3 January 2002, Write Stuff Enterprises.
29. Edgar Northrup, interview by Richard F. Hubbard, tape recording, 9 January 2002, Write Stuff Enterprises.
30. Ray, interview, 24 May 2001.
31. Ibid.
32. Pergam-Ball, interview.
33. Ray, interview, 24 May 2001.

Chapter One Sidebar

1. Pergam-Ball, interview.
2. Knauf, interview.

Chapter Two

1. Ray, interview, 24 May 2001.
2. David Porreca, interview by Richard F. Hubbard, tape recording, 24 May 2001, Write Stuff Enterprises.
3. Karl Gould, interview by Richard F. Hubbard, tape recording, 19 October 2001, Write Stuff Enterprises.
4. Ray, interview, 24 May 2001.
5. Ibid.
6. Ray Staszewski, *San Diego Union*, 12 June 1983.
7. Porreca, interview.
8. Staszewski.
9. Ibid.
10. *San Diego Corporate Guide to Business* (1985).
11. The Titan Corporation 1984 Annual Report.
12. Staszewski.
13. Ibid.
14. Porreca, interview.
15. *San Diego Corporate Guide to Business* (1985).
16. Staszewski.
17. Titan 1984 Annual Report.
18. Titan 1984 Annual Report; *San Diego Corporate Guide to Business* (1985).

19. Titan 1984 Annual Report.
20. Gary Loda, interview by Richard F. Hubbard, tape recording, 9 October 2001, Write Stuff Enterprises.
21. *San Diego Corporate Guide to Business* (1985).
22. Ibid.
23. Ibid.
24. Ibid.
25. Titan 1984 Annual Report.
26. Ibid.
27. Ibid.
28. Ibid.
29. Ibid.
30. Ibid.
31. Ibid.
32. Staszewski.
33. Titan 1984 Annual Report.
34. Ibid.
35. Ibid.

Chapter Three

1. Dan Berger, "Defense Spending Increase Is Good for S.D.," *San Diego Union*, 28 January 1985.
2. Ray, interview, 24 May 2001.
3. Ibid.
4. Rolf Erikson, interview by Richard F. Hubbard, tape recording, 9 October 2001, Write Stuff Enterprises.
5. Ray, interview, 24 May 2001.
6. McDougall, interview.
7. Ibid.
8. Dan Berger, "Newly Merged Titan Hits the Deck Running," *San Diego Union*, 30 May 1985.
9. John Arme, interview by Richard F. Hubbard, tape recording, 29 November 2001, Write Stuff Enterprises.
10. Gene Ray, interview by Jeffrey L. Rodengen, tape recording, 23 May 2001, Write Stuff Enterprises.
11. Berger, "Newly Merged Titan."
12. William M. Alpert, *Barron's*, 6 January 1986.
13. Ray, interview, 23 May 2001.
14. Julie Flynn and Dave Griffiths, *Business Week*, 24 August 1987.
15. Ibid.
16. Alpert.
17. Ibid.
18. Ibid.
19. Ibid.
20. Ray, interview, 24 May 2001.
21. Ibid.
22. Ray, interview, 23 May 2001.
23. Janet Lowe, *San Diego Tribune*, 10 June 1986.
24. Ibid.
25. Janet Lowe, "Investment Watch," *San Diego Tribune*, 11 December 1985.
26. Karl Sussman, *The Stock Market Magazine*, July 1986.
27. "Titan-Corp; TSES receives IVDS award from Honeywell," Business Wire, 27 January 1986.
28. Flynn and Griffiths.
29. John Kerr, *Electronic Business*, November 1986.
30. "Antimissile Aide to Work for Military Contractor," *New York Times*, Associated Press, 31 July 1986.
31. Kerr.
32. Bill Zettinger, interview by Jeffrey L. Rodengen, tape recording, 9 October 2001, Write Stuff Enterprises.
33. Bill Olsen, "Titan Corp. Is on a Buying Binge," *San Diego Business Journal*, 3 March 1986.
34. Ted Kavanaugh, interview by Richard F. Hubbard, tape recording, 10 October 2001, Write Stuff Enterprises.
35. Ray, interview, 24 May 2001.
36. Erikson, interview.
37. *Los Angeles Times*, 18 June 1985.
38. Marshall Nelson, interview by Richard F. Hubbard, tape recording, 6 December 2001, Write Stuff Enterprises.
39. Business Wire, 6 October 1986.
40. Business Wire, 28 June 1985.
41. Alpert.
42. Ray, interview, 24 May 2001.
43. The Titan Corporation 1986 Annual Report.
44. Alpert.
45. *San Diego Union*, 2 November 1986.
46. Kavanaugh, interview.

Chapter Three Sidebar

1. Nancy Jenkins, interview by Jeffrey L. Rodengen, tape recording, 4 December 2001, Write Stuff Enterprises.
2. Philip Spence, interview by Richard F. Hubbard, tape recording, 3 December 2001, Write Stuff Enterprises; Brian Williams, interview by Richard F. Hubbard, tape recording, 7 December 2001, Write Stuff Enterprises.

Chapter Four

1. Chris Kraul, "Titan Corp. Determined to Pull Itself out of Doldrums," *Los Angeles Times*, 21 April 1987.
2. Ibid.
3. Andy Gaspar, interview by Jeffrey L. Rodengen, tape recording, 3 January 2002, Write Stuff Enterprises.
4. Gene Ray, interview by Jeffrey L. Rodengen, tape recording, 9 October 2001, Write Stuff Enterprises.
5. Kraul, "Titan Determined to Pull Itself Out."
6. Ibid.
7. Knauf, interview.
8. The Titan Corporation 1987 Annual Report.
9. David Coburn, "Despite Defense Cuts, High-Tech Firms Predict Modest Growth," *San Diego Tribune*, 29 January 1988.
10. Ibid.

11. Ibid.
12. Business Wire, 3 April 1987.
13. Business Wire, 23 April 1987.
14. Business Wire, 27 July 1987.
15. Business Wire, 2 July 1987.
16. Business Wire, 10 December 1987.
17. PR Newswire, 22 November 1988.
18. S. Lynne Walker, "Titan Vice President Quits to Rejoin Sandia Labs," *San Diego Union*, 10 May 1989.
19. David Coburn "Titan Hunting a Niche," *San Diego Tribune*, 13 May 1998.
20. Ibid.
21. Ibid.
22. Business Wire, 3 August 1987.
23. PR Newswire, 3 January 1989.
24. Anne Middleton, "Titan Sells Meteor-Technology Unit to Employees," *San Diego Business Journal*, 12 June 1989.
25. Ibid.
26. Business Wire, 31 March 1987.
27. Donald Coleman, "Titan Pressing Ahead on Small Acquisitions," *San Diego Tribune*, 13 May 1987.
28. Ray, interview, 23 May 2001.
29. Coleman, "Titan Pressing Ahead."
30. David Coburn, "C³I=Hope for Defense Contractors," *San Diego Tribune*, 10 October 1988.
31. Ibid.
32. Ibid.
33. S. Lynne Walker, "Titan Is Taking Aim at Conventional Weapons," *San Diego Union*, 13 May 1988.
34. *San Diego Union-Tribune*, 20 January 1989.
35. *San Diego Union-Tribune*, 18 February 1989.
36. PR Newswire, 2 February 1989.
37. PR Newswire, 2 August 1989.
38. Ibid.
39. PR Newswire, 12 September 1989.
40. PR Newswire, 27 September 1989.
41. "Good Times Over, Defense Firms Fear," *San Diego Union-Tribune*, 29 January 1990.
42. Ibid.

Chapter Five

1. Associated Press, 24 August 2000.
2. Ibid.
3. S. Lynne Walker, "If Pentagon Bleeds, Suppliers Feel Pain, *San Diego Union*, 29 January 1990.
4. Ibid.
5. Knauf, interview.
6. Donald C. Bauder, "Analyst Says Titan in Strong Position," *San Diego Union*, 25 February 1990.
7. Ibid.
8. Robert Hanley, "Job Losses Shadow Proposed Defense Dept. Budget Cuts," *San Diego Tribune*, 28 April 1990.
9. The Titan Corporation 1990 Annual Report.
10. Ibid.
11. "Pentagon on Search for Key Spares, Despite Stockpiles," *Aerospace Daily*, 24 August 1990.
12. Tom Trimble, interview by Richard F. Hubbard, tape recording, 10 October 2001, Write Stuff Enterprises.
13. Ibid.
14. McDougall, interview.
15. Zettinger, interview.
16. Michael Kulinski, interview by Richard F. Hubbard, tape recording, 23 May 2001, Write Stuff Enterprises.
17. Kent Redding, *San Diego Daily Transcript*, 17 May 1991.
18. Craig Rose, *San Diego Union*, 13 March 1991.
19. *Software Industry Report*, 19 August 1991.
20. PR Newswire, 21 January 1992.
21. Gene Ray, *San Diego Union*, 27 August 1991.
22. Ibid.
23. Diane Lindquist, *San Diego Union*, 30 October 1991.
24. Ibid.
25. PR Newswire, 23 September 1993.
26. Ray Calhoun, interview by Richard F. Hubbard, tape recording, 9 October 2001, Write Stuff Enterprises.
27. PR Newswire, 2 June 1992.
28. Tom Allen, interview by Richard F. Hubbard, tape recording, 23 May 2001, Write Stuff Enterprises.
29. Ibid.
30. *San Diego Daily Transcript*, 23 January 1995.
31. Elizabeth Douglass, *San Diego Union-Tribune*, May 21, 1993.
32. The Titan Corporation 1992 Annual Report.
33. Libby Brydolf, "Economic Conversion: Technology's Not the Problem," *San Diego Business Journal*, 21 December 1992.
34. Ibid.
35. Daniel J. Fink, interview by Richard F. Hubbard, 30 November 2001, Write Stuff Enterprises.
36. McDougall, interview.
37. Titan 1992 Annual Report.

Chapter Six

1. Ken Kreyenhagen, interview by Richard F. Hubbard, tape recording, 3 December 2001, Write Stuff Enterprises.
2. PR Newswire, 17 June 1992.
3. Libby Brydolf, *San Diego Business Journal*, 17 May 1993.
4. Ibid.
5. Ibid.
6. Elizabeth Douglass, "For Titan, Quick Shift to Civilian Products," *San Diego Union-Tribune*, 19 April 1994.

7. The Titan Corporation 1993 Annual Report.
8. Business Wire, 27 April 1993.
9. Elizabeth Douglass, *San Diego Union-Tribune*, 8 March 1994.
10. The Titan Corporation 1994 Annual Report.
11. Elizabeth Douglass, *San Diego Union-Tribune*, 24 May 1993.
12. Titan 1994 Annual Report.
13. Douglass, "For Titan, Quick Shift," 19 April 1994.
14. McDougall, interview.
15. PR Newswire, 8 February 1994.
16. Douglass, "For Titan, Quick Shift," 19 April 1994.
17. Titan 1994 Annual Report.
18. Ibid.
19. Douglass, "For Titan, Quick Shift," 19 April 1994.
20. Ibid.
21. Ibid.
22. Ray, interview, 24 May 2001.
23. Bruce V. Bigelow, *San Diego Union-Tribune*, 31 January 1995.
24. Ibid.
25. Ibid.
26. Titan 1994 Annual Report.
27. Bruce V. Bigelow, *San Diego Union-Tribune*, 26 December 1995.
28. Elizabeth Douglass, *San Diego Union-Tribune*, 22 November 1995.

Chapter Seven

1. Bruce V. Bigelow, "Titan Plans Bond Offering to Raise Cash," *San Diego Union-Tribune*, 19 September 1996.
2. Business Wire, 13 November 1996.
3. "Titan to Help Army Tap into Commercial Satellites," *Defense Daily*, 11 January 1996.
4. PR Newswire, 13 March 1996.
5. Kulinski, interview.
6. Ibid.
7. Ibid.
8. PR Newswire, July 1996.
9. PR Newswire, 4 September 1996.
10. PR Newswire, 29 July 1996.
11. Bigelow, "Titan Plans Bond Offering."
12. Dan Gallagher, "SAIC and Titan Get Lucrative Federal Contracts," *San Diego Daily Transcript*, 3 January 1997.
13. Bigelow, "Titan Plans Bond Offering," 19 September 1996.
14. Bob Starzynski, "Titan Corp. Charts Bold Acquisitions Course," *Newsbytes*, 18 September 1998.
15. M. C. "Bud" Baird, interview by Jeffrey L. Rodengen, tape recording, 24 May 2001, Write Stuff Enterprises.
16. Eric DeMarco, interview by Richard F. Hubbard, tape recording, 9 October 2001, Write Stuff Enterprises.
17. Ibid.
18. Ray, interview, 24 May 2001.
19. PR Newswire, 6 May 1997.
20. PR Newswire, 13 February 1997.
21. "Titan Announces Spin-Off of Internet Software Business to Private Investment Group," PR Newswire, 19 May 1997.
22. Virginia Oliver, interview by Richard F. Hubbard, tape recording, 23 May 2001, Write Stuff Enterprises.
23. "The Titan Corporation Is Member of Alcatel-Led Consortium Selected by PT Multi Media Asia Indonesia to Develop Innovative Asian Multimedia Satellite Network," PR Newswire, 14 July 1997.
24. Martyn Williams, "Telecom Roundup-Titan Corp Joins Asian Satellite Consortium," *Newsbytes*, 15 July 1997.
25. PR Newswire, 18 February 1997.
26. PR Newswire, 1 July 1997.
27. Bruce V. Bigelow, "IPOs Get Short Shrift Lately; Titan's Offering Not the Only One," *San Diego Union-Tribune*, 20 June 1998.
28. "The Titan Corporation Announces Agreement with PointCast to Provide Corporate Broadcast Solutions," PR Newswire, 19 September 1997.
29. "The Titan Corporation Partners with Janus Technologies to Provide End-to-End Information Technology Asset Management Solutions," PR Newswire, 5 November 1997.
30. Allen, interview.
31. "Titan Announces Launch of First Turnkey Medical Device Sterilization System for In-House Manufacturing Use," PR Newswire, 7 February 1997.
32. Denny Olson, interview by Richard F. Hubbard, tape recording, 9 October 2001, Write Stuff Enterprises.
33. The Titan Corporation 1999 Annual Report.
34. Williams, interview.
35. Allen, interview.
36. The Titan Corporation 1997 Annual Report.
37. Ibid.

Chapter Eight

1. Starzynski, "Titan Corp. Charts."
2. "Titan Subsidiary Licenses Needle Technology to Influence," *Medical Industry Today*, 26 March 1998.
3. "Titan Wins Contract to Fix Wyoming's Y2K Problems, PR Newswire, 11 September 1998.
4. *EDP Weekly*, 9 March 1998.
5. Earl Pontius, interview by Richard F. Hubbard, tape recording, 2 May 2002, Write Stuff Enterprises.

6. Baird, interview, 24 May 2001.
7. Starzynski, "Titan Corp. Charts."
8. Ibid.
9. Diane Scott, interview by Richard F. Hubbard, tape recording, 10 December 2001, Write Stuff Enterprises.
10. Kreyenhagen, interview.
11. Bruce V. Bigelow, *San Diego Union-Tribune*, 2 July 1998.
12. PR Newswire, 13 January 1999.
13. Porreca, interview.
14. "Titan Corporation Acquires Atlantic Aerospace Electronics Corporation," PR Newswire, 26 July 1999.
15. Baird, interview, 24 May 2001.
16. Kulinski, interview.
17. Ibid.
18. Porreca, interview.
19. Ibid.
20. Oliver, interview.
21. Porreca, interview.
22. Titan 1999 Annual Report.
23. Andrea Siedsma, *San Diego Business Journal*, 26 April 1999.
24. Allen, interview.
25. Bruce V. Bigelow, *San Diego Union-Tribune*, 16 December 1999.
26. Siedsma, *San Diego Business Journal*.
27. Larry Oberkfell, interview by Richard F. Hubbard, tape recording, 9 October 2001, Write Stuff Enterprises.
28. Ibid.
29. Titan 1999 Annual Report.
30. Oberkfell, interview.
31. Olson, interview.
32. Allen, interview.
33. Oberkfell, interview.
34. Titan 1999 Annual Report.

Chapter Nine

1. Nick Wakeman, *Washington Technology*, 10 January 2000.
2. Ibid.
3. Ibid.
4. PR Newswire, 1 February 2000.
5. PR Newswire, 15 March 2000.
6. Nick Wakeman, "Titan Joins Big Boys," *Washington Technology*, 3 April 2000.
7. Ibid.
8. Baird, interview, 24 May 2001.
9. Chuck Saffell, interview by Richard F. Hubbard, tape recording, 3 May 2002, Write Stuff Enterprises.
10. The Titan Corporation 2000 Annual Report.
11. Dean Calbreath, "Titan Slashes Projections; Stock Drops," *San Diego Union-Tribune*, 11 April 2001.
12. Ashley Dunn, *Los Angeles Times*, 31 August 2000.
13. Dean Calbreath, *San Diego Union-Tribune*, 18 August 2000.
14. Mike Allen, "Titan Buys Datron," *San Diego Business Journal*, 13 August 2001.
15. Ed Bersoff, interview by Richard Hubbard, tape recording, 17 December 2001, Write Stuff Enterprises.
16. Ibid.
17. Ibid.
18. Don Bauder, "Titan Deal for BTG Helps Both," *San Diego Union-Tribune*, 9 December 2001.
19. Baird, interview, 24 May 2001.
20. Ibid.
21. Saffell, interview.
22. Ibid.
23. Ibid.
24. Joseph Saponaro, interview by Richard F. Hubbard, tape recording, 2 May 2002, Write Stuff Enterprises.
25. Ibid.
26. BTG, "Technology with a Purpose," promotional material.
27. BTG Delta Research Division, "Parametric Cost Engineering for Schools," promotional material.
28. BTG, "Geographical Information Systems," promotional material.
29. Earl Pontius biographical sketch, Titan Systems Corporation.
30. Pontius, interview.
31. Ibid.
32. Ibid.
33. Ibid.
34. Ibid.
35. Ibid.
36. Saffell, interview.
37. Ibid.
38. Ibid.
39. Ibid.
40. Ibid.
41. Ibid.
42. Ron Gorda, interview by Richard F. Hubbard, ape recording, 3 January 2002, Write Stuff Enterprises.
43. M. C. "Bud" Baird, interview by Richard F. Hubbard, tape recording, 6 February 2002, Write Stuff Enterprises.
44. Tony Frederickson, interview by Richard F. Hubbard, tape recording, 6 February 2002, Write Stuff Enterprises.
45. Ibid.
46. Ibid.
47. Ibid.
48. Ibid.
49. Ibid.
50. Spence, interview.
51. Frederickson, interview.
52. Ibid.
53. Ibid.
54. Ibid.
55. Gorda, interview.
56. Ibid.
57. Ron Jacobson, interview by Jeffrey L. Rodengen, tape recording, 23 May 2001, Write Stuff Enterprises.
58. Ibid.
59. Ibid.
60. Gorda, interview.

61. "Titan Awarded U.S. Army Contract to Provide 83 Prophet Production Systems," PR Newswire, 18 June 2001.
62. The Titan Corporation 2001 Annual Report.
63. Gorda, interview.
64. Ibid.
65. Andrea Siedsma, *San Diego Business Journal*, 27 March 2000.
66. "Cayenta's Mainsaver EAM Selected by Plant Services as 'Problem Solver Awards' Winner," PR Newswire, 22 October 2001.
67. "Cayenta's e.System to Power Express.com's Operations," PR Newswire, 15 December 2000.
68. Titan 2000 Annual Report.
69. Kulinski, interview.
70. Titan 2001 Annual Report.
71. Ibid.
72. "E-tenna's New AccuWave Product Line Enhances GPS Antenna Accuracy," PR Newswire, 19 June 2001.
73. "E-tenna Corporation, a Titan Subsidiary, Opens New State-of-the-Art Antenna Testing Facility," Business Wire, 29 May 2001.
74. "E-tenna Corporation and Ashvattha Semiconductor Form Alliance to Accelerate Realization of Software Defined Radios," Canadian Corporate Newswire, 25 September 2001.
75. Titan 2001 Annual Report.
76. "SureBeam Wins 10th Patent, Further Advancing Its Position in Electronic Pasteurization Technology," Business Wire, 4 June 2001.
77. Dean Calbreath, "A Taste of Irradiation, SureBeam Gets Exotic Fruit to Mainland U.S.," *San Diego Union-Tribune*, 19 October 2001.
78. Loda, interview.
79. Oberkfell, interview.
80. Ibid.
81. Ibid.
82. Ibid.
83. Titan 2001 Annual Report.
84. Loda, interview.
85. "Titan Creates Homeland Security Office Focused on Chemical and Biological Terrorism,' PR Newswire, 17 October 2001.
86. "Dr. Gene Ray of Titan Corporation," *NBC News Transcripts*, 29 October 2001.
87. Information provided by Ralph "Wil" Williams, 31 May 2002.
88. "Bioterrorism Scare Heightens Concern Over Safety of Food Supply," PR Newswire, 8 November 2001.
89. Loda, interview.
90. Ibid.
91. Ibid.
92. Bruce V. Bigelow, "Titan Corp. to Buy Jaycor for $95 Million in Stock," *San Diego Union-Tribune*, 23 January 2002; "Titan Acquisition of Jaycor, Inc. Closes," PR Newswire, 21 March 2002.
93. "Titan Acquisition of GlobalNet Closes," PR Newswire, 21 March 2002; "Titan to Acquire GlobalNet," PR Newswire, 7 January 2002.
94. Mike Allen, "Titan Announces Albuquerque Firm as Third Acquisition of the Year," *San Diego Business Journal*, 4 March 2002; Bruce V. Bigelow, "Titan Corp. to Acquire Company in New Mexico," *San Diego Union-Tribune*, 26 February 2002.
95. Titan 2000 Annual Report.
96. DeMarco, interview.
97. Spence, interview.
98. Zettinger, interview.
99. Jerry Grossman, "2002 May Be Banner Year for Govt. IT Public Offerings," *Washington Technology*, 7 January 2002.
100. Rochelle Bold, interview by Richard F. Hubbard, tape recording, 10 October 2001, Write Stuff Enterprises.
101. Titan 2001 Annual Report.
102. Gorda, interview.
103. Ray, interview, 24 May 2001.
104. Fink, interview.
105. Joseph Caligiuri, interview by Richard F. Hubbard, tape recording, 30 November 2001, Write Stuff Enterprises.
106. Mary Squazzo, interview by Richard F. Hubbard, tape recording, 24 May 2001, Write Stuff Enterprises.
107. Gorda, interview.
108. Ray, interview, 24 May 2001.

Chapter Nine Sidebar

1. Congressman Randy "Duke" Cunningham, interview by Richard F. Hubbard, tape recording, 5 December 2001, Write Stuff Enterprises.
2. Congressman Duncan Hunter, interview by Richard F. Hubbard, tape recording, 5 December 2001, Write Stuff Enterprises.

INDEX

Page numbers in italics indicate photographs.

A

ABC News, 125
accelerators, 61
AccuWave, 122
acquisitions and mergers
    ACS (Advanced
        Communications
        Systems), 102
    ADS (Advanced Digital
        Systems), 49
    Afronetwork, 88
    APA (Applied Power
        Associates), 123
    ARAP (Aeronautical Research
        Associates), 39
    Atlantic Aerospace Electronics
        Corp., 93
    AverStar, 102–103
    Beta Development Corp., 26
    BTG Inc., 104
    California Computing
        Resources, 40
    California Research and
        Technology, 91
    Compunet, 27
    Contech Media, 120
    CRT (California Research and
        Technology), 40
    Datron Systems Inc., 104
    DBA Systems, 89
    Delfin Systems, 89–90
    Diversified Controls Systems, 78
    DSC (Defense Systems Corp.), 39–40
    Eldyne, 78
    Gateway System, 120
    Horizons Technology, 89
    Jaycor, 128
    J. B. Systems, 95
    LinCom Corp., 102
    Linkabit Corp., 57–58
    Mainsaver, 95
    Meteor Communications, 40
    PAI (Physics Applications Inc.), 49
    PSI (Pulse Sciences Inc.), 49
    Pulse Engineering, 102
    SenCom, 102
    Solutions for Growth, 95
    Spectron Development
        Laboratories, 40
    SRC (System Resources
        Corp.), 92–93
    Stonehouse Group, 59
    TNP (Transnational Partners), 92
    Unidyne, 78
    Validity Corp., 89
    VisiCom Laboratories, 89
acquisition strategy, 39, 49–50, 90–91
ACS (Advanced Communications Systems), 102
Adaptive Maneuvering Logic, 26
ADS (Advanced Digital Systems), 49
Advanced Products & Design Division, 112
Aerospace Corp., 14
Aerospace Daily, 56
Aerospace Electronics Division, 112
Affiliated Computer Services, 103
Afronetwork, 88
AFWAL (U.S. Air Force Wright Aeronautical Laboratories), 40
A. G. Edwards, 102
airport security, *51*, 52, 60
Air Traffic Systems Division, 109
Alcatel Telspace, 81
Alexander, Michael, 103, *129*, 130
Allen, Charles, *129*
Allen, Tom, 61, 84, 96
Allen-Bradley, 40
Allied Signal, 36, 39
AMC (Artificial Magnetic Conductor), 121–122
American Electronics Association, 60
American Precision Plastics, 88
analyst opinions
    on 2000 acquisitions, 102–103
    bullish on Titan, 55–56
    on commercial ventures, 69–70
    on defense industry slowdown, 46
    on divestitures, 1987-1989, 49
    on Titan-EMM merger, 34–36, 43

INDEX

Anchor Foods, 125
anthrax-tainted mail, 126, *126*
APA (Applied Power Associates), 123
APCO Project 25 radio, 117
Applied Engineering Solutions Group, 107–108
Applied Technologies Group, 113–114
ARAP (Aeronautical Research Associates), 39
Arme, John, 32
Arthur Andersen and Co., 32, 80
Ashvattha Semiconductor, 122
Atlantic Aerospace Electronics Corp., 93
Atlantic-based Fleet Technical Support Center, 79
AverStar, 102–103
Aviation Engineering Group, 111
AWACS (Airborne Warning and Control System) radar, 67, *67*, 109
awards and recognition
    Association of Corporate Growth, 51
    Cogswell Award, 41
    MRO Problem Solvers Award, 117

B

B-1B bomber, *33*
B-52s, *10*, 16, *16–17*, 79
Babbitt, Albert, 18, 39
Baird, Melton C. "Bud," 80, 89, 93, 103, *103*, 104–105, 112
Bank for Agriculture and Agricultural Cooperatives, 77
Bank of America, 12
Barron's magazine, 34, 43
Baxter Healthcare, 84
Bell Atlantic, 78
Berger, Duaine, *43*
Bersoff, Edward H. "Ed," 104, 130
Beta Development Corp., 26, *26–27*, 28, 60
Beyster, J. Robert "Bob," 16, 20
BICC (Battlefield Intelligence Collection Capability), 114
BKP Capital Management, 73
Bletzacker, Frank, *42*
Bloomberg News, 125

board of directors, *35*, 35–37, 39, *39*, 103, *129*, 130
Bold, Rochelle, 130
Boles, Knop & Co., 102
Bowyer, Royce, *43*
Branch, Al, 112
Brennan, Tom, 111
B. Riley & Co., 117
Brookhaven National Laboratory, 59
Brown, Harold, 31
BSE-Mediscan, 72
BTG Inc., 104
Bunker Ramo, 37
Bureau of Labor Statistics, 107
Bush, George, 55

C

C^2 (command and control), 89
C^3I (command, control, communications, and intelligence), 24, 24–25, *25*, 27, *28*, 47, 51, 52, 79
C^4ISR (command, control, communications, computers, intelligence, surveillance, reconnaissance), 103, 104–105, *105*, 109, 110, *110*
Calhoun, Ray, 60–61
California Computing Resources, 40
California Research and Technology, 40, 91
Caligiuri, Joseph, *35*, *129*, 130
Cap Gemini America LLC, 88
Cargill, 96, 123
Cayenta, 23, 94–96, *95*, 117, 117–119, *118*, *119*
Cayenta Assist, 118–119
Cayenta e.System, 117–118
CellularVision Technology & Telecommunications, 77–78
Century Electric, 40
cesium 137, 60–61
Cheney, Dick, 55–56
Chittagong Stock Exchange, 120
CIA, 109
Cipher Data Products, 36
Civil Government Services Group, 107–108
Civilian Space Division, 107
Civil Sector, 107–108

CNN, 127
CNNfn, 101
Cohen, Peter, 130
Cold War, 23, 25, 45, 53, 55, 60
Comarco, 59, 79
Commerce Business Daily, 13
commercial applications of Titan products, 47, 52, 56, 60, 62–63, 87, 105. *See also* SureBeam technology
commercial contracts, 48, 53, 59, 72, 88
Commissariat a l'Energie Atomique, 88
Communications and Software Solutions Division, 117
Communications Engineering Group, 111
Communications Products Division. *See* Linkabit
Compunet, 27
Computer Electronics & Marketing Association, 68
Computer Power Products, 42
Computer Sciences, 103
COMSEC/Vocoder, *115*
Conner, Dave, 79, 111
Conning International, 39
Contech Media, 120
Cooper, Robert, 93, 112
Cornerstone Promotions, 118
corporate culture, 21, 28–29, 38, *38*, *41*, 91–92, 128–129
Corporate Guide, 25
corporate strategy
    acquisitions, 39, 49–50, 90–91
    diversification, 47, 62–63
    divestitures, 120–121
    early years, 12, 21, 24, 28–29
    spin-offs and spin-outs, 72, 80–81
    See also corporate culture
Cruttenden and Co., 55–56
CT&T (CellularVision Technology & Telecommunications), 72
Cubic Corp., 43, 55, 67
Cunningham, Randy "Duke," 122

D

DAMA (demand assigned multiple access), 57, 67, 69, 76–77, 93, *114*, 114–115. *See also* Mini-DAMA

DAMALink, 69
DARPA (Defense Advanced Research Projects Agency), 40
Datron Advanced Technologies Division, 116
Datron Systems, *101*, 104
Datron World Communications, 116
DBA Systems, 89
Defense & Intelligence Systems Group, 109
Defense Communications Agency, 47
defense contracts, 42, 47–48, 52–53, 59, 61–62, 67, 75–76, 78–79, 84
defense industry, 29, 36, 46, 55–56, 71
Defense Intelligence Agency, 110
Defense Logistics Agency, 109
Defense Nuclear Agency, 48
Defense Systems segment, 78
Delfin Systems, 80, 89–90
Del Monte, 125
Delta Data Systems, 67
DeMarco, Eric, 75, 80–81, 128, *128, 129,* 130
Department of Commerce, 53
Department of Energy, 56, 60, 114
Department of Health and Human Services, 107
Department of Transportation, 109
Derbyshire, Harry, *35*
Dillon Read & Co., 36
diversification, 47, 62–63. *See also* commercial applications of Titan products; SureBeam technology
Diversified Applications Group, 112
Diversified Controls Systems, 78
divestitures, 35
    Advanced Materials, 49
    Army training simulation business, 67
    Canada Alloy Castings, 42, 48–49
    Canada Investment Castings, 48–49
    CAST, 49
    Computer Power Products, 42–43
    Indiana General Ferrite Products, 40
    Indiana General Motors Products, 40
    Meteor Communications, 49
    SCI Systems electronics division, 78
    television encryption business, 75
Donoghue, Dan, 111
Doppler radar systems, *44*
Dowe, Mike, 26
downsizing, 48–49
Driessen, Kenneth, 41, 46
Drug Enforcement Administration, 56
DSC (Defense Systems Corp.), 39–40
DST (Direct Solution of Turbulence), *52,* 53
DynCorp Aerospace Technology, 76

E

E-2C aircraft, *66, 68*
E-3 klystrons, *67*
Eldyne, 78
ElectriCities, 118
electron beam technology, 60–61, 74. *See also* SureBeam technology
electron guns, *26–27*
electro-optics, 26
Emerging Technologies segment, 78
Emmpak, 96
EMM-Titan merger, 31–36, *38,* 40
employee benefits, 29, 128–129
employees, *22, 34, 38, 91*
Engineering and Information Services segment, 106–112
Environmental Protection Agency (EPA), 107
Erikson, Rolf, 31–33, *32,* 40
E-tenna, *121,* 121–122
Excel, 125, *125*
EXDEP (explosive detection with energetic photons), 51, 52
Executive Information System, 82–83
explosion-detection technology, *60*
Explosive Standoff Minefield Breacher Demonstration and Validation program, 72
Express.com, 118

F

F-15 aircraft, 41–42
F-16 aircraft, *37, 48*
FAA (Federal Aviation Administration), 48, 82–83, *83,* 93, 109, 118
Federal Communication Commission, 107
Federal Deposit Insurance Corp., 107
Feibusch, Robert J., 73
Feibusch & Co., 73
Ferris Baker, Watts, 87
Financial and Regulatory Division, 108
financial data
    1986, 43
    1988, 52
    1993, 66
    1998, 87, 92
    1982–1984, 28
    1990–1992, 62
Fink, Daniel, *35,* 63, *129,* 130
five-year anniversary, *37*
Florida Power & Light, 117
FOG-M (Fiber Optic Guided Missile), 48
food irradiation, 26–27, *74,* 84–85, 96–99, *124,* 125
Frederickson, Tony, 71, 113
Fricke, Martin, 43
Frost & Sullivan, 75

G

Gaspar, Andy, 31, 46
Gateway System, 120
GBS Joint Venture, 58
General Dynamics, 12, 55, 71
General Electric, 36
General Electric Aerospace, 62
General Instruments, 65
Gogolak, Victor, 12, 20
Golding, Susan, 126, *129,* 130
Gorbachev, Mikhail, 45
Gorda, Ronald B., 71, 75, 112, 114–115, 130–131
Gould, Karl, 23, 32
Grossman, Jerry, 102
Grossman, Steven, 122
Ground Based Sensor, 58
GSM cellular network, 119–120
GTE, 78
GTE Sylvania, 13, 20

Guardian radio, *101*
Guidant Corp., 84, 88
Gutenstein, Robert, 65, 70

H

Hahn, David A., 71
Hanisee, Robert, 35, *129*, 130
Hard Mobile Launcher, 27
Hawaii Pride, 98
Hay, Roger, 80
headquarters of Titan, 29, *29*
heart-rate monitor, noncontact, 28
Henry, Patrick, 122–123
Hensel, Neil, 71
Hercules AN/AAR-47 missile warning system, 61
Hercules Defense Electronics, 61
Homeland Security Office, 126
Horan, James J., 36
Horizons Technology, 12, 18, 89
Houlihan Lokey Howard & Zukin, 102
Houston Satellite Systems, 65
HTI/DNA, 20
Hughes Aircraft, 36
Huisken Meats, 96, 123, 124, *124*
Hunter, Duncan, 122

I

IBM, 18, 20, 39
IBP, 96, 123, *125*
ICMS (Integrated Cargo Management Systems), 66
ImagClear Model F5000 High Volume Fingerprint Card Scanner, 89
IMED Corp., 72
Indiana General Ferrite Products, 40
Indiana General Motors Products, 40
Indosat, 82
Information & Logistics Support Division, 109
Information Products Group, 114–117
Information Security Systems Division (ISSD), 114
Information Solutions Group, 89, 109
initial public offerings, 31–32, *100*, 125
Integrated Installation & Engineering Group, 111–112

Integrated Services Division, 111
Intelliscape, 112
international business, 76–77, 81–82, 88, 94, 119–121, *121*
INTERPRETER, 26
iSBC single-board computers, 41–42
IT Outsourcing Division, 107
IT Services & Solutions, 107–108
ITT Federal Services, 59

J

Jacobs, Irwin Mark, 57
Jacobson, Ron, 115
Janus Technologies, 83
Jaycor, 128
J. B. Systems, 95
Jenkins, Nancy, 38
Joint STARS program, *92*, 109
joint ventures and alliances
    Ashvattha Semiconductor, 122
    GBS Joint Venture, 58
    Hawaii Pride, 98
    Janus Technologies, 83
    Motorola, 67
    Motorola Government Electronics Group, 58
    PointCast, 83
    PT Multi-Media Asia Indonesia, 81–82
    Sakon, 88
    Soliance Network, 95
    SureBeam Brazil, 124
    Texas A&M University, 124
    Titan-Delta Data Systems, 67
    Titan-ICMS, 66
    Titan-MPI, 66
    Titan-Patriotic Scientific Corp., 66
    Titan Satellite Systems Corp., 65
    Zero Mountain, 99
Jordan, Mark, 102, 104
Joseph Charles & Associates, 80
Judge, Frederick, 68

K

Kalb, Voorhis & Co., 65, 70
Kavanaugh, Ted, 39, *39*
Keahl, Gerry, 39
Kearney, Mary, 24, 25, 38, *38*

Keenan, Linda Frady, 20
Kelly, Todd, 39
Klein, Mark, 96
klystron test facility, 67
Knauf, Albert E. "Ed," 20, *32, 41*
    early career, 12, 14–15, 17–18
    founding of Titan, 11–13
    military career, 14–15
    named executive vice president, 41
    resignation, 71
    at SAIC, 17–18
    Titan, naming of, 18
Knauf, Sue, 11
Knop, Richard, 102
Konchan, Tom, 88
Kraft Foods, 123
Kreyenhagen, Ken, 65, 71, 91
Kulinski, Mike, 76–77, 93–94
Kutler, Jon, 70

L

LaBlanc, Robert, *129*, 130
lawsuits, 66, 72–73, 103–104
Leedom, Bud, 78
licensing agreements
    British Aerospace Dynamic Group Bristol Division, 41
    Influence Inc., 87
    Intel, 41
    Messerschmitt-Bolkow-Blohm, 60
Lieser, Tom K., 71
LinCom Corp., 102, 122–123
Lindsey, William C., 102, 123
Linkabit (also called Linkabit Wireless, Titan Linkabit, Communications Products Division), 57–58, 66–67, 71–72, 75, 78–79, 80–82, *82*, 93, 114
Linkabit Corp., 57–58
Linkabit smart card, 65–66
Litton-PRC, 103
Llewellyn, Richard, 12
Lloyd-Butler, Thomas, 46
Lockheed Corp., 31
Lockheed Martin, 58, 103
Loda, Gary, 26–27, 60, 124, 126–127
Logicon, 103
logos, *19, 31, 96*
Los Angeles Society of Financial Analysts, 68

Los Angeles Times, 12
Loxley Public Co., 72
LSM-1000 modems, 76
LST-5 Manpack radios, 67, 72, 82, 116, *116*

M

M²A (Multi-Media Asia Satellite Telecommunications System), 81
M/A-COM, 57
MAGTF (Marine Air Ground Task Force), 112
mail irradiation, 126–127
Mainsaver, 95, 119
Manpack. *See* LST-5 Manpack radios
Mainsaver EAM (Enterprise Asset Management), 117, 119
MainsaverNow, 117
Mantech, 59
Maritime Sector, 111–112
Martin, Julie, *40*
Martin Marietta Energy Systems, 47
McClellan Air Force Base, 67, *67*
McDougall, John R. "Jack," *20*, 24, *32*, 41
   on diversification, 63, 68
   EMM-Titan merger, role in, 32
   founding of Titan, 11–13
   on Linkabit, 57
   resignation, 71
MCI, 78
McKewon & Timmins, 43
Meagher, Thomas, 87
Medical Industry Today, 87
medical sterilization, 61, *88*
MEMS (Micro Electro Mechanical Systems), 115
Metaxas, John, 101
Meteor Communications, 40
Meyer, Stephen P., 71
military aircraft, *10, 16, 33, 37, 42, 48, 66, 68, 79, 82, 92, 105, 110*
Milstar satellite communications, 62, 63, 76
Mini-DAMA (demand assigned multiple access), 57, *66*, 66–67, *68*, 78–79, 82, *82*, 114–115
Minuteman missiles, *18*, 20
Motorola, 58, 67, 72, 82

MPI (Message Processing International), 66–67
MRO Marketplace, 117
MULTIBUS computer products, 78
Multi-Missile Fire control unit, 48
MunicipalityManager, 119
MX missile system, 13, *13*, 20

N

naming of Titan, 11, 18
NASA, 56, 107, 109, 117
NASDAQ, *100*, 125
National Imagery and Mapping Agency, 109
National Imagery Organization, 110
National Laboratory of Frascati, Italy, 59
National Reconnaissance Office (NRO), 110, 112
National Security Advisory, 114
National Security Agency, 110
NATO, 77, 114
NATO Maritime Command and Control Information System, 111
Naval Ocean Systems Center, 52
Naval Research, Office of, 53
Naval Research Laboratory, 114
Naval Surface Warfare Center, 52–53
Naval Training Systems Center, 47
Naval Undersea Warfare Center Division, 84
Naval Weapons Support Center, 48
Navy Fleet Satellite Communications Systems, 52
Nelson, D. Marshall, 38–40
New York Stock Exchange (NYSE), 32, *129*
New York Times, 37
NEXRAD (Next Generation Weather Radar), 53, *56*, 65
Nippon Iron Powder, 49
Northrup, Edgar, Col., 16, 18, 21
NovAlert emergency notification system, 67
NRG, 117
NRO. *See* National Reconnaissance Office
nuclear power generating station, San Onofre, Calif., 27
Nynex, 78

O

OADS (Optical Air Data System), 48
Oberkfell, Larry, 97, *123*, 123–125
Oliver, Virginia, 95–96
Olson, Denny, 84, 98
Omaha Steaks International, 125
OpenGL display graphics, 117
Operation Desert Shield, 56, 58
Operation Desert Storm, 58
Operations, Analysis & Training Group, 110
Operations Security Fundamentals, 108
Osterloh, Bob, 109

P

Pacific Missile Test Center at Point Mugu, Calif., 59
Pageau, Gary, 61
PAI (Physics Applications Inc.), 49
Pasifik Satelit Nusantara, 72, 76–77, 82
Patent and Trademark Office, 107
Patriotic Scientific Corp., 66
Pentagon, *15*
Pergam-Ball, Cathy, 12, 19, 21
Perry, Bill, 31
Physics International, 37
Pilots Associates Program, 40
Plant Services, 117
PointCast, 83
Pontius, Earl, 89, 109–111
Porreca, David, 23–24, 92, 94–96, 117, 119
Porter Novelli, 127
Pratt, Kathy, 65
ProfitMAX, 66
profit sharing, 21, 29
Programmable Display Generator, 37
Prophet, 93, 115–116, *116*, 130
Pryt, Robert K., 73
PSI (Pulse Sciences Inc.), 49, 114
PSN. *See* Pasifik Satelit Nusantara
PT Multi-Media Asia Indonesia, 81–82
publicity
   CNNfn interview, 101
   Internet rumors, 103–104
   regarding mail irradiation, 125–127
   See also analyst opinions

INDEX

Pulse Engineering, 102
Pulse Sciences. *See* PSI
pulse technology, *59*

Q, R

Quarterdeck Investment Partners, 70
Ramo, Simon, *39*
Rapid Retargeting Technology, 93
Ray, Gene W., *11, 14, 23, 30, 32, 35, 37, 38, 49, 57, 102, 129, 131*
   congressmen's comments about, 122
   corporate strategy, 39, 56, 62, 70, 79–81, 87
   early career, 14–18
   founding of Titan, 11–13, 18
   management style, 91–92, 128–129, 130–131
   media appearances, 101, 125–127
   recruitment of employees, 23–24
   on Soviet Union disbanding, 60
   Titan-EMM merger, 31–33, 35
   youth, 13–14
Raytheon, 48, 82
Reagan, Ronald, 23, 25, 45
recruiting employees, 21, 23–24, 37–39
RESAL Saudi Corp., 125
Research and Technology Division, 113–114
restructuring
   1984, 26
   1987, 48–49
   1995, 71
   1996–1997, 79–81
   1999, 93
   2000, 121
Rochialle Corp., 88
Rockwell International, 31, 40
Rohr Industries, 36, 55
Rose, Les, 107
Roth, Jim, *129*
Royal Australian Air Force, 82
rumors affecting Titan stock, 103–104
RUPS (rotary uninterruptible power supply), 42, *43*

S

Saab Aircraft, 59
Safeguard Scientifics, 90
Saffell, Chuck, 103, 105–107, 111–112
SAIC (Science Applications International Corp.), 11–12, 16–17, 20, 39, 103
Sakon, 88, 120
Samma Microwave, 59
Sandia Laboratories, 52
San Diego, defense industry in, 29, 36, 55
San Diego Business Journal, 49, 62, 96
San Diego Police Dept., 28
San Diego Stock Report, 78
San Diego Union-Tribune (formerly *San Diego Union, Evening Tribune*), 24, 33, 51, 55, 56, 68, 71, 75, 78, 104, 124
Saponaro, Joseph, 103, 107
Sara Lee, 124
Satellite Business News, 65
satellite communications products, *45, 57*, 58, *58, 64*, 69, 75–77, *77*, 93, *94*, 119–121
Satimo test chamber, 122
Scherman, Bob, 65–66
Schwan's, 125
Science, Technology and Products Segment, 112–117
SCIS Food Services, 125
Scott, Diane, 91
SDI (Strategic Defense Initiative), 25, 25–27, 31, 37, 45, 55
SEA (Science & Engineering Associates), 128
Secretary of Defense Crisis Coordination Group, 47
SECS (Severe Environment Computer System), 61, *62*
Selwitz, Larry, 55–56
Sempra Energy, 95
SenCom, 102, 109
sense of humor (Titan people), 38, *38, 41*
September 11, 2001, terrorist attacks, 108, 113
Series 200 Mask Repair Station, 28
SESCO (Severe Environment Systems Co.), 33, 41

Shipboard Communications and Computer Access Terminal, 52
Short Range Air Defense System, 27
Siam Bank of Thailand, 72
SIGINT (Signal Intelligence), 114
single channel transponder system, *57*
Sithe Energies, 118–119
Small ICBM program, 27
Soliance Network, 95
Solutions for Growth, 95
Sopp, Mark, 128
Soviet Union, 45, 60
Space and Electronic Warfare Systems Command, 111
Space and Naval Warfare Systems Command, 52
SPD (Signal Products Division), 115
Spectron Development Laboratories, 40
Spence, Phil, 114, 129
Squazzo, Mary, 129–130
SRC System Resources Corp., 92–93
staff of Titan, *22, 34, 38, 91*
Star Wars missile defense system. *See* SDI (Strategic Defense Initiative)
stock performance, 75, *98*, 99, 101, 103–104
Stonehouse Group, 59
Strategic Defense Command, U.S. Army, 47
STRATUS computers, 28
Strike Command, 110
SureBeam technology
   development of, 26–28, 60–61, 74
   food irradiation, 84–85, 88, 96–99, 123–125, *124, 125,* 127
   logo, *96*
   mail irradiation, 125–126
   medical use of, 61, 72, 84, 88, *88*, 123
   military uses of, 127–128
   SureBeam Brazil, 124
   turnkey systems, *84*, 84–85, 88
Systems and Imagery Division (formerly DBA Systems), 117
Systems Engineering Division, 114

Systems Integration Division, 114
Systems Management Services Division, 109
Systems Services Group, 112

T

TACTICIAN, 26
takeover attempt, 72–73
Tech Ion Industrial, 124
Technical Resources Sector, 109–111
TEMPEST products, 77
Test and Evaluation Division, Information Solutions Group, 109
Test Program Sets, 79
Texas A&M University, 124
Thermo Electron Corp., 90
Thierry, Lauren, 101
Titan Applied Technology Group, 71
Titan Beta, 38
Titan Broadband, 120
Titan-Delta Data Systems, 67
Titan Electronics, 41, 59
Titan-EMM merger, 31–36, *38*, 40
Titan Engineering Services, 59
Titan-ICMS, 66
Titan Information Services, 60
Titan Information Systems, 68–70, 72, 77
Titan Linkabit. *See* Linkabit
Titan-MPI, 66
Titan-Patriotic Scientific Corp., 66
Titan Research & Technology, 71, 73
Titan Satellite Systems Corp., 65
Titan Scan, 61, 72, 73, 96–99, 123
Titan Software Systems, 78, 80, 82–83, 92
Titan Sterilization and Pasteurization Systems, 80, 84–85, 88
Titan Systems (original company), 11–13, 19–33
Titan Systems (subsidiary of Titan Corp.), 33, 41, 71, 93, 101–107
Titan Systems Technical Resources Sector, 89
Titan Technologies, 41
Titan Technologies and Information Systems, 80, 83–84, 89–90

Titan Wireless, 93–94, 119–121, *121*
TNP (Transnational Partners), 92
Tomahawk cruise missile, *28*
TomoTherapeutics, 87
Tracor Aerospace, 72
Trident submarine, *52*
Trimble, Tom, 57
TRW, 20, 34, 35, 36, 55
TT&C (telemetry, tracking and control) systems, 116
Tyson, 96, 123

U

UCOM (United Communication Industry Public Co.), 77
UHF Tunable Patch Antenna, 93
UMTS (Universal Maintenance Training Systems), 47
Unidyne, 78–79, 111
United States Surgical Corp., 88
U.S. Air Force, 15–17, 28, 67, 76, 109
U.S. Air Force Crisis Management centers, 47
U.S. Air Force Worldwide Military Command and Control Information System Program, 47–48. *See also* WWMCCS
U.S. Army, 72, 75, 89, 109, 110
U.S. Army Communications and Electronics Command, 75–76
U.S. Army Medical Research Center, 28
U.S. Coast Guard, 109
U.S. Customs Service, 89
U.S. Defense Information Systems Agency, 109
U.S. Marine Corps, 47, 72, 110, 111
U.S. Navy, 66, 67, 76, 78, *78*, 79, 84, 111, 112
U.S. Postal Service, 107, 108, 126, *126*
USA Today, 125, 127
USS *Cole*, *113*
UtilityManager, 117, 119

V

Validity Corp., 89
Valleylab, 88
Value Line, 35

video encryption technology, 65–66, 69
Video PassPort, 69, *70*, 72, 77–78
video-teleconferencing system, *24*
Virgile, Lucien, 35–36
VisiCom Laboratories, 89
Viterbi, Andrew, 57
VLSI (very large scale integration) chips, 55
VME computer products, 78, 115
VSAT (very small aperture terminals), 77, 119

W

Warburg Pincus, 31, 46
Washington Technology, 90, 103, 130
Webb, J. Sidney "Sid," 30, 32, 34, 49, 57
    on acquisition strategy, 49–50
    annual report 1990, 56
    career, 33–34
    sense of humor, 38
    Titan-EMM merger, role in, 31–33
Weinstein, David, 80
Wenaas, Eric, 128
WGSA, 20
Williams, Brian, 61, 85
Wilshere, Ken, 110
Woolsey, R. James, 35, 39, *39*
Wright, Joseph R., Jr., *129*, 130
WW Johnson Meat Co., 125
WWMCCS (Worldwide Military Command and Control System), 18
Wyoming, state of, 88

X, Y, Z

Xpress Connection, 76
X-Ray Needle, 87
Y2K compliance services, 83, 88
Yonas, Gerold, 37, *39*, 41, 52
Young, Henry, *42*
Zahn, Paula, 127
Zante, Tony, 60
Zero Mountain, 99
Zettinger, Bill, 39, 58, 129